PAPER TEXTILES

Christina Leitner

PAPER TEXTILES

A&C Black London

The author

Christina Leitner studied textiles with psychology, philosophy and education at the Mozarteum University in Salzburg. She wrote her final year dissertation on the history of paper textiles in different cultures. The Austrian Department of Education awarded her a grant which enabled her to continue her studies at the Institute for Textiles, Art and Design at the Arts University Linz.
In terms of practical work, Christina Leitner has been working intensively with different paper yarns for the past few years. She has been looking at the different ways in which they can be used in textile techniques, in particular in weaving. She is currently working as a lecturer at the Arts University Linz where she gives various courses and lectures on textile-related topics.

We would like to thank *Woodnotes Oy* and *Finnish Paper Yarn Ltd*. for their kind support.

Front cover
Dress: Gisela Progin
Background: Veronika Rauchenstein

Back cover
Above left: Veronika Rauchenstein

All other motifs on the cover: Christina Leitner

First published in Great Britain 2005
A&C Black Publishers
37 Soho Square
London W1D 3QZ
www.acblack.com

ISBN-10: 0 7136 7444 X
ISBN-13: 978 0 7136 7444 6

Published in German 2005 by
Haupt Verlag
Falkenplatz 14
3001 Bern
Switzerland

ISBN 3-258-06767-8

© 2005 Haupt Bern

A CIP record for this book is available from the British Library

Editor: Regine Balmer
Designed by Atelier Muhlberg, Basel
Printed in Germany by Passavia, Passau

Contents

Preface

'When he has carried enough silk, he changes to paper'

Japanese saying

Paper: whether cut into strips or twisted into threads and made into material – this trivial, yet amazing idea has fascinated me for years. The clash of paper textiles in different cultures has drawn me into an exciting world, where everything is a continuous learning process. I was always discovering new facets of the expressive, flexible paper yarns and meeting people who have devoted their lives passionately to this material. Today, paper strings offer a contemporary means of expression with great creative potential. However, the possibilities that paper has to offer as a textile are still only known to a small group of people.

Paper strings have been part of our daily lives for a long time, whether as handles on shopping bags or for tying up parcels, but we never really notice them. Interior design shops have now begun to sell different variations of table sets, baskets, mats and lampshades made from paper yarn; plaited summer hats and woven bags are now often seen in the collections of many fashion houses. When it is dyed with different colours, the yarn appears similar to raffia and hardly anyone notices that the item is made up of 100% paper! If this is specified, then the products tend to elicit an element of surprise. Even though paper strings can be found everywhere today, they still attract very little attention.

The complexity involved in making paper bears witness to the wealth of invention which has taken place over the years. Instead of twisting loose fibres directly into threads, these are, via paper as a "diversion", made into a non-woven fleece. Finally, after a long and involved process, they are transformed into yarn. Paper strings are made from a highly cultivated material which has the effect of being both natural and immediate, which is what makes it so appealing.

I have concentrated on two completely different types of paper yarn in this book. I am fascinated by the compact, industrially-produced strings which have been available here in various colours and strengths for several years now. The technology used to manufacture them was invented during wartime. At this time, paper yarns were widely distributed as a cheap substitute for the raw textile material which was sorely lacking. Once the war was over, however, they soon faded back into obscurity. It was only in the 1980s in Finland that people began to appreciate the strange beauty of this material once again. The extravagant, yet simple rugs and accessories from the North quickly gained international recognition. This has led to textile designers experiencing a huge increase in demand for products made from paper strings in the last few years.

In addition to learning about this robust material, I also became acquainted with a second, less well-known type of paper yarn, with a far longer history. The Japanese had been using homemade paper for centuries. They would produce the yarns by hand and work these into the most exquisite garments. This material was called Shifu. As was the case in the West, this process was driven by necessity. However, over time, Shifu conquered even the highest echelons of society and became a cult object in itself, embued with deep symbolic meaning. This tradition has an exciting history and experts outside of Japan have been looking into the impressive technique for several years now. Once Shifu has worked its magic on you, it never leaves.

Because of different cultural backgrounds, production methods and raw materials, the character of European and Japanese paper yarn is very different. However, the basic principle, that of obtaining a textile-like material, is the same. If we examine the two more closely, we can see that there are many other parallels between the two cultures. This is how I came up with the idea of attempting to bridge the gap by looking at both areas in one book. The precious knowledge which has been handed down and the few documents and articles which are still available and which experts have generously allowed me to use, should together provide a comprehensive overview of these topics. In many respects, this book is an attempt to bridge the gap between the two areas. I have linked their historical roots to current developments in paper textiles to provide in-depth background information on the topic.

This book should also be seen as a practical reference guide and will, I hope, awaken the reader's interest in creating their own designs. I discuss both Japanese Shifu yarns and European paper strings in this book and I describe how they can be used with different textile techniques. You will find concrete, detailed instructions along with further ideas which are designed to inspire your own creations. The 'Future' sections, together with the Gallery section, demonstrate paper's flexibility as a textile and the different areas in which it has been used.

I would like to thank all those who have, in any way, made a contribution to this book. Those who have lent me their knowledge, their materials and their time during my research. Those who have allowed me to use their practical work or pictures and who have welcomed me so warmly on my travels. I have greatly enjoyed being a part of the enthusiasm of so many people and getting to know new things about the material through them. From soft, filigree creations made from fibres as thin as air to sturdy rugs made from thick cords. My meetings with passionate artists, impressive craftspeople and industry designers were as varied as the material itself. Without the colour that so many people have brought to it, this book would not be as it is today. My intention then, using the widest possible palette, is to show how multi-faceted and flexible this material is, no matter who is working with it. To this end, the book is also a reflection of current developments and is intended, however indirectly, to give something back, where I have received so much.

I would like to thank the Swiss weaver, Mäti Müller, in particular. She has been there from the very start of this project, provided me with contacts and was extremely helpful when I needed expert advice. I first learnt about Shifu from her and she helped fire my passion. I would also like to thank Akiko Sato for so sensitively acquainting me with Japan, her home.

Writing a book is a wonderful, yet demanding task – nobody knows this better than my family! I would like to thank with all my heart Roswitha, Erich, Monika and Thomas for their active support and loving understanding.

Introduction: Paper as textile

A life without paper is unimaginable these days. Culture, as we understand it, would be almost impossible without paper. As the medium on which we write, it represents memory, human progress and the dissemination of knowledge. In other documents, it creates identities and authorities. As money, it regulates the comparison of products and services. But even aside from its significance as a platform for words or pictures, it plays an important role today as a material in packaging, hygiene and entertainment products. The modern paper industry produces over 3000 different types of paper. It has to provide for, on average, a usage of 200kg per year per person. In addition to its cultural value, paper therefore also stands for short-term usage, versatility and consumption.

In the age of electronic information processing, the significance of the traditional description matter would, at first glance, appear to be diminishing. But now, more than ever, as we question how necessary paper is as the only possible carrier for knowledge and culture, we are beginning to discover many new qualities. Its sensual and haptic qualities are at the centre of attention. Softness, hardness, the surface structure, colour and acoustic qualities as well as being able to reuse it after recycling; these have all become important criteria. In the last 35 years, paper has also moved into the field of applied art as an independent material for expression.

Paper has, without doubt, many, varied uses. In this book, there will be an introduction to paper in one of its most sophisticated forms, i.e. as fibres and threads, and how they are turned into textiles. Some surprising things about paper will also be uncovered. As the qualities of the finished fibres always have a direct connection to the source material, we first need to find out what paper actually is. In this way, we are nearer to finding out more about the basic nature of paper fibres themselves.

What is Paper?

The history of human development is marked by people's need to describe their clashes with nature in pictures and written symbols. So, in a tradition which spanned thousands of years, people continued to invent new and costly material on which to express this. In ancient Egypt, for example, layers of fibre were laid across each other crosswise. They were then hammered and pressed into *papyrus*. In Asia Minor and Europe, however, animal skins were being made into *pergament*. It must have been in the 2nd century AD, when the invention was made (most probably in Mongolia) which would continue to influence human culture and which was to go on to conquer many different continents after Asia: That invention was paper.

The East
A Chinaman, Ts'ai Lun, is said to be the founder of paper. He was the first person to undertake systematic research in this field and, in 105 AD, he wrote down his findings about the new material. Whilst looking for a new material on which to write, Ts'ai Lun experimented with textile waste and plant fibres. The key reform was not in trying to connect the fibre bundle by hammering or pressing it together, but rather by connecting the single base fibre to a compact surface in a *creative process*.

To make paper you need *cellulose fibres* which are found mainly in the bark, stalks, leaves or seed pods of almost all plants. However, the quality and the composition of these vary. These fibres are soaked, boiled and mashed until there are just single cells left. When they are stirred into cold, and as soft and clean as possible, water, you get a glutinous mass: the *pulp* or *suspension*. This is then put into a large container, known as the *tub*. During the papermaking process, the fibre pulp is put onto a porous, sieve-like surface. While the cellulose fibres are held by the sieve, the superfluous water drains off and a compact fibrous structure remains in the sieve. It acquires an automatic inner stability from the natural cohesion of the fibres (this occurs on drainage). With no outside intervention, a piece of paper is born!

Paper production was a strictly kept secret for a long time in China. Nevertheless, the wisdom spread slowly across Asia, crossing Korea and reaching *Japan* in approximately AD610. The process was improved there and it was not long before Japan developed an incredibly high paper culture which has not been equalled to this day.

As a new source material for extracting cellulose, the Japanese used homegrown plants, primarily the bast of the paper mulberry bush which had particularly long and stable fibres. They also added a root extract to the fibre pulp, giving the pulp an ideal consistency and making it possible to create much thinner, yet regular, and more stable sheets. This newly-developed immersion process using a rolling sieve went on to revolutionise the papermaking process.

For hundreds of years this work-intensive craft was a winter occupation for the vast majority of Japanese farmers and was an integral part of family life. Paper was not only used to write or draw on, but also for many daily and religious uses. There was never a hard and fast set of rules for making it. The process changed according to the consistency of the fibres, temperature, humidity and what it was to be used for. A certain feel for the material was therefore necessary. After industrialisation, the number of papermakers dropped considerably. However, there are still a few workshops in Japan which continue to make top-quality paper by hand following the over 1400-year old tradition.

The West
The spread of paper to the West goes back to the 8th century. Arabian armies brought Chinese prisoners of war home with them and these prisoners knew how to make paper. The Chinese prisoners of war were forced to share their knowledge. As the plants they needed did not exist in the surrounding areas, they experimented with textile remains from old bits of clothing (rags and cloth) and achieved amazing results in a very short time. The new paper was compact and firm and soon became an important trade item. Paper reached Egypt and Morocco at the beginning of the 12th century, over a thousand years after its initial discovery; and finally reached Europe, later going on to America in the 17th century.

From the very beginning, Western paper production was different to the traditional Eastern method in many ways. Many large paper mills were built along the river in many towns. Specialised craftsmen worked there in very hard conditions. To make the fibre pulp, lumps of linen, cotton and hemp were used. Initially, they had to be cut up by hand until 1670 when the so-called *Dutchman* was invented. This was a machine with rotating knives, and it took over this hard task. Water was added to the fibre pulp and the resulting *pulp* was put into a large, wooden tub at which the papermaker worked. He used a papermaking sieve, which was made up of two separable parts. The base of the sieve was a wooden frame on which a sieve enclosed by metal wires was fixed. There was also a separate lid in the form of a loose frame which covered the edge of the sieve from above. Both parts were dipped into the pulp and fibre pulp absorbed. Once the water had drained off, the papermaker took the lid off and gave the bottom section of the sieve containing the non-woven fleece to the so-called *coucher*. His task was to put the wet sheet onto a pile (*couching*) and give the mould back to the the papermaker. The individual sheets were separated with woolen felt to prevent them sticking. Other craftsmen then had to press the pile to drain the water out, hang the sheets up to dry and then, according to how they would be used, either smooth them out or spread them with glue.

This craft form hardly changed at all over the centuries. The material was seen to be very expensive and for a long time it was only accessible to aristocrats or church dignitaries. But when printing was discovered in the 15th century, the demand for paper increased rapidly. This led to a huge shortfall in raw materials, as textile waste was only available to a limited extent. So an industrious search for suitable substitutes began. Finally, in 1719, the French scientist, René-Antoine Réaumus, struck gold: he discovered the secret of the Canadian wasp. It builds its nest by chewing splinters of wood, glues them together with its saliva and builds the foundations of its nest with the resulting, paper-like substance. This is how broken down wooden fibres came to be the new base material for making paper.

As time went on, many technical changes have made the work-intensive process much easier. The most far-reaching reform was, without a doubt, the invention of the paper machine at the end of the 19th century. It made making paper by hand unnecessary and laid the foundations for the modern paper industry. The introduction of mechanical and chemical processes with which wood could be made into *wood pulp* or *cellulose*, also made papermaking much easier. These brought with them an almost endless variety of types of paper and cardboard which could be produced. This is how paper as a material changed from being a valuable craft product to a cheap mass-produced item over the last two hundred years. However, even today, handmade Japanese paper is still seen as very precious and it cannot be equalled in strength, shine and durability.

How much of a textile is paper

Paper and textiles have an exciting and constantly changing relationship. In terms of their structure, qualities and usage, they have many similarities. On closer examination, it is even harder to make a clear distinction between them. Even in the definition of paper, there are terms which are quite 'textile-like':

"Paper is a two-dimensional material. It is made from fibres which come primarily from plants and is moulded in a sieve by draining the swollen fibrous material. This gives a fibrous felt which is then compressed." (from: Trobas, Karl: '*abc des Papiers, Die Kunst Papier zu machen,*' Graz 1982)

When he says "fibres which come from plants", this means cellulose. This also forms the basis of numerous other textiles such as cotton, linen or viscose. As they share the same raw substance, these materials share many qualities. During the ageing process, they encounter similar problems and this is particularly important in restoration work.

Even the way in which the paper fibres are joined together is similar to the processes used with other textiles. The natural adhesion of the fibres to one another gives the surface its strength. This is why people often talk of 'felting' or of a 'non-woven fleece' in papermaking. The process is linked to felting, but if we examine it more closely, we can

see that there is a big structural difference: Natural fibres which are used for producing textiles almost always have a minimum length of 5mm (0.2 in.) and are in bundles. If we take the felting of wool as an example, it only acquires its inner cohesion by removing the flaky outer layer of the fibres, leaving every single fibre whole. Paper, on the other hand, has a different inner structure. Chemical combinations also stabilise the fibrous framework. This is what happens: during the preparation of the pulp, the fibres are boiled and beaten, thus changing the composition of the cellulose. The fibre walls are at least partially destroyed so that the smaller basic units, the so-called *fibrils*, are freed. If these broken-down fibres are mixed with water, they are capable of absorbing water molecules. Then, in the draining and drying processes, they deposit the water molecules, thus binding the individual parts closer together. This process is known as *hydration*. It makes the cellulose fibres undergo new chemical combinations – an essential effect for making paper. You are left with a compact non-woven fleece whose strength has not been produced externally. The individual fibres are usually no longer visible to the naked eye and generally appear laid crosswise in the finished sheet, giving a relatively high level of stability. Therefore, whilst mechanical processes are usually required to make natural textiles; chemical combinations take place when making paper.

The close relationship between paper and textiles is also clear from the similar ways they are put to use: Numerous examples in the history of many cultures indicate that reinforced and oiled paper was used as a textile, for example as cloaks, floor coverings or sails. Traces of these examples can still be found in parts of Asia, the Philippines, Korea and Japan.

The precursor to paper provided remarkable proof of how similar the two groups of material are. Bark raffia, known as *Tapa* or *Amate*, was used for clothing, interior furnishing textiles, to write or draw on or for ceremonial purposes for thousands of years. It was used in many different countries the length and breadth of the Equator. A structure made from bark was used. The bark was boiled and pieces were placed on top of one another like a lattice. Then the fibrous layers were beaten with special wood hammers until they joined together, doubled in width and became a supple, cohesive construction. In contrast to the relatively stiff and very fragile papyrus which is made in a similar way, bark raffia is very flexible and elastic. As fabric, it is painted or printed, making it very textile-like.

If we take a closer look at how paper is used today in our culture, the hazy border between the two materials becomes even clearer: During the 20th century, paper has taken the place of textiles in many areas. Handkerchiefs, napkins, handtowels, nappies, bags, tablecloths, toilet paper, coffee filters, teabags – we are familiar with all these paper objects today. They have similar textile-like qualities to their predecessors, but are much cheaper and more practical. We can just throw them away after using them once so they fit in perfectly with our time-poor lifestyles. Today, *non-woven fabrics* are often used. The technology for these was developed in the 1960s. These include non-woven, synthetically-produced fleeces which have both a paper and textile-like effect. Using various methods, the fibres are entangled and made into a two-dimensional surface by sticking, melting and other similar processes. Cheap clothing, bed linen and accessories for the home made of non-woven fleece were popular in the 1960s and 1970s. Today the material is still used as padding and for hygienic reasons in hospitals, the hospitality industry, scientific laboratories and in the military. For theatrical costumes and for decorative purposes, a material known as *Tyvek* was used. In the last few years, its paper-like character has been used in everyday clothing. We have come full circle: paper is used as a substitute for textiles to make our daily usage of it more simple. But also, people have recently begun to imitate the look of paper using high-tech materials to transport up-to-date contents. This interplay between textile and paper is continuous. It is difficult to pinpoint any one moment where paper and textiles separate as there are many similarities, but there also structural differences. It is noticeable however, that since papermaking by hand increased in the 1960s and as the art of paper slowly established itself, many textile workers began to tackle the new material. At ease working with fibres and needing to 'create' from the qualities of a material, the characteristics of paper seemed to appeal in particular to those who made textiles. Today there are

paper workshops in many university textile clases and the art of paper is included in academic writing on textiles.

Paper can therefore be seen as a textile, even if we are not talking about textiles in the classic sense. But when paper is cut into strips, twisted into a yarn and finished using many varied textile techniques, we are left in no doubt: paper definitely has a textile side!

Tapa material is made of beaten raffia bark and is thought to be
one of the precursors to paper. It has been used as a textile for
thousands of years.

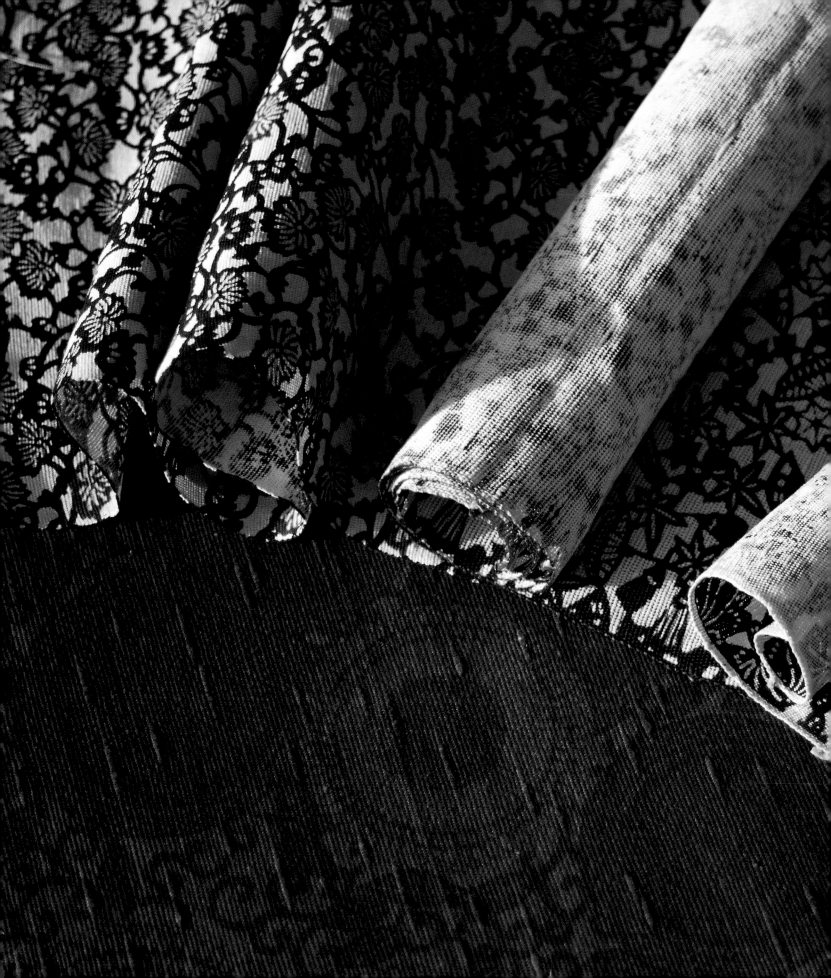

1 History

▲ Damask weave by Dora Jung from Finland 100% paper, about 1950 (with thanks to the Helsinki Design Museum

Stencil painted Japanese shifu weave made from 100%
◄ paper, about 1940

1.1   Paper textiles in ancient Japan

Japanese aesthetics and paper culture

The Japanese are well-known for picking up on external impulses and perfecting them in such a way that they develop a completely independent dynamic and are able to give the world something 'typically Japanese'. If we take paper, which was actually invented in China, as an example, we can see this phenomenon even more clearly. Because of its geographic and political isolation, Japan became very independent on a cultural level. This is still in evidence in many areas today, even despite Western influence.

It is difficult to compare the art of Japanese papermaking with Western paper culture. From a tradition over 1400 years old, it produced a product which far exceeds our concept of paper as a commodity. The expression '*washi*', which is used generically in the West to describe all handmade Japanese paper, actually covers an entire philosophy behind the production process and use of paper. It is virtually untranslatable.

In Japan, paper has always been closely associated with spirituality, culture and natural experience. This is obvious from the Japanese language: *kami* is the Japanese expression for an actual piece of paper, but has the shared meaning of 'divinity'. Although these have different origins, it is nonetheless clear that there is a close, emotional relationship between these two areas.

In ancient Japan art and craft were inextricably linked to religion and meditation. Shintoism is the original religion of the Japanese. Its basic idea is that all substances, whether people, animals, plants or objects, are inhabited by gods (*kami*). They fill the world with life and are reborn through nature or through the actions of people. This belief that all things were living inevitably had a huge effect on how people related to their environment. To a large extent, they expressed themselves by respecting nature and there was the craftsman who not only created living objects, but 'built new housing for the

gods'. He had to listen to the language of the material. His self-realisation was less important than referencing nature.

Many shintoist elements remained even after the introduction of Buddhism in AD538. This spiritual religion advises taking a simple, meditative path in order to reach spiritual enlightenment. Paper was honoured as the mediator between God and the people. These religious teachings were not considered to be contradictory. Rather, over the course of time, they merged into a specific, Japanese spiritual way of thinking.

Paper is seen as a godly material and has become an indispensable symbol for religion in Japan. It plays an important role as cord, typewriter ribbon,

Washi, handmade paper, has a special significance in Japanese culture.

flag, folded shapes, suits or talismen. The production process itself is almost a spiritual act. The papermaker must be pure in himself and understand the process intellectually. This is so he is able to produce a high-quality product as paper has always been seen as a mirror of the soul. It is difficult to make and easy to destroy or soil. However, if treated well, it can have an extraordinarily long life. In this way it has extremely 'human' qualities and becomes a metaphor for life and death.

The colour white stands for purity and viriginity in Asian culture. It also stands for sorrow and cosmic influence. In this way paper became an absolute symbol of culture and human development. The combination of spiritual and creative principles has always been central to Japanese aesthetics. In the West, people are used to thinking in terms of difference. The Japanese, on the other hand, believe that all opposite pairs such as creativity and spirituality, tradition and progress, craft and technology, functionality and aesthetics, craft and art, West and East, Buddhism and Shintoisim, being and not-being etc. are just two sides of a self-contained unity. They can both be the key to good creation, but what makes a difference is stripping it all down to the essential. This stems from the core belief that big things are born of small things. In the West we perceive complete structures and external forms first, we assess building and basic conditions and only look at the detail later.

The original Asian way of seeing, on the other hand, works the other way round – they see the detail, then the whole. In all areas of Japanese art, individual modules and units play an important role as building blocks for the complete structure. This can be seen, for example, in the measurements of the rice straw floor mat, the basis for all the proportions in the Japanese house. The kimono is made only made from right angles with equal proportions. Even the learning process for a craft follows this principle. By watching intensively and repeating the same steps, the whole process can be understood and internalised, just as it can with papermaking.

In Japan, paper was used in numerous aspects of daily and religious life since it was first discovered. In the West, however, until the 20th century, paper was seen almost exclusively as a medium for writing or drawing. In Japan, it has always been seen as a means of expression in its own right.

In the Heian age (AD794–AD1185), it was fashionable, among poets for instance, to have special papers made by a personal papermaker. These papers served as a source of inspiration for poetry and were intended to emphasise the emotional content. The poem could not be folded any old how, it could only be used in combination with the correct paper as a self-contained whole. Papermakers and poets contributed equally to this complete work of art.

But even aside from writing and art, *washi* became central to almost all areas of life. In addition to the many religious and ritual purposes it served, it was also used to make lanterns, shades, pockets, kites, toys, masks and packaging. In the traditional Japanese house, paper has become an essential building material. Because of their receptiveness to nature and the special light that they spread, the famous paper windows (*shoji*) and blinds (*fusuma*) have significantly contributed to the unique character of Japanese architecture. It is therefore not surprising that paper has also been used to make clothes for hundreds of years in Japan.

Kamiko

Kamiko translates as 'paper shirt'. This word is made up of *kami* (paper) and *koromo* (monk's robe) because, according to legend, a Buddhist monk called Shoku invented the paper shirt. In AD988, he apparently used the pages of old *Sutras*, the holy scripts of the Buddha, to make himself a provisional choir shirt when he was expecting guests and had run out of clean clothes.

Visitors were impressed by the monk's modesty and the intensity of his religious beliefs. By wearing the holy scripts next to his skin, he really seemed to want to internalise them and demonstrated that he knew he was safe in them. This is how this almost ritualistic form of body-covering spread amongst communities of monks and then into other areas of the population.

Even though the legend did not report this, it does show the close connection between kamiko and spiritual values. The fact is that the paper dress has a history of over 1000 years in Japan and, over the years, it has been worn by people in many different social classes.

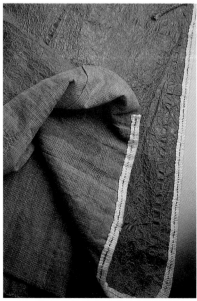

Detail of a historical riding coat (*jibaori*) made of kamiko. It has been waterproofed with *kakishibu* starch sap. This is how it gets its typical rust colour.

Simple kamiko jacket from the beginning of the Edo age.

This Japanese ink drawing shows the various workstations involved in the process of making kamiko.

Production process and qualities

Kamiko is making clothing out of paper which is treated in a special way, to increase its fabric-like character. Fairly thick washi is used for this purpose. This is usually made from *kozo*, the fibres of the paper mulberry bush. The papers are waterproofed with natural substances and creased following a set system. The corners are folded inwards and the sheet is made into a ball. It is then rubbed and crumpled until it is soft and supple. This process is repeated several times and the paper is smoothed and waterproofed again on each occasion. This produces a product which appears very similar to a textile. It has a crease-like structure and is called *momi-gami* (creased paper).The momi-gami is very resistant and flexible. It is water, wind and heat-resistant. In ancient Japan it was not made only into clothing, but it was also used in other daily aspects of Japanese culture.

Depending on the plant extract used for the waterproofing, the colour of the paper changes. In this way, over time, two different types of kamiko material have developed. Under-privileged people tended to wear paper dresses which were painted with tannic acid obtained from ebony plants. This was a persimmon sap, known as *shibu* or *kakishibu* and it coloured the paper ochre yellow to rust brown. Kamiko material which is to stay white, was painted with the starchy sap, *konnyaku*. This is extracted from the *Amorphophallus konjac*, an amanita plant from the family of the aron plant. *Konnyaku* makes the paper last much longer without changing its natural colour. White kamiko was only worn by monks, and later members of the nobility.

After creasing and waterproofing, the individual sheets of paper are glued together with plant paste to make long strips. Clothes were then cut from the balls and glued or sewn together to make the end product. The garments kept people warm because of the isolating effect of the paper. However, as kamiko could not be washed, the paper dresses were always seen as something provisional and this only served to emphasise their mystical effect.

Cultural and historical development

The origins of kamiko probably date back to the impoverished rural population. In the Heian age (AD794-1185), social relations were very difficult and farmers continually lacked the necessary raw products to make their textiles. They usually made paper themselves in the winter months, so this material was used for many things. It has been reported that rural workers used oiled and stained paper to protect plants in the field from frost. One farmer was apparently once caught out in the rain whilst working in the field, and to protect himself, he pulled the sheets over him. He was later thought to have initiated the making of clothes from kamiko. Out of necessity, farmers were certainly among the first people to make use of the advantages of the waterproofed paper garments.

Other population groups also began to use the material for clothes and thereby gave it a completely new meaning. Buddhist monks produced their kamiko themselves in an almost meditative process. They saw it as a symbol for their religious beliefs. Because the paper clothes simply fell apart after they had been worn a few times, they were seen as a metaphor for the transitoriness of life and the cycles of nature. The creased kamiko also had a timeless effect and when worn, it was not possible to stand out from others or to show off in it as it was not very flattering to the figure. In this way, kamiko perfectly embodied the Buddhist ideals of simplicity and reduction.

Artists and travelling poets prized the humourous element of kamiko clothing above all else. Whoever wore paper rustled when they walked and looked

Embossed and painted kamiko bags and paper from the workshop of
Mashiko and Tadao Endo, a famous Japanese papermaking couple
who produce kamiko and shifu.

rather clumsy and awkward. They made it clear in this way that they did not take themselves too seriously and could laugh at life's absurdities. It has been passed on that poets also used the provisional clothes as writing material so that they could keep their spontaneous literary outpourings close to their bodies. In numerous *haiku*, the famous Japanese short poems, kamiko is used as a metaphor for material poverty and spiritual wealth. A famous poet in the Edo age (AD1615-1868) wrote, 'As I lie dying now, who on earth should I leave my kamiko to?' An old, used kamiko had little or no value, yet this 'exquisite poverty' had great spiritual significance.

Somewhat later, in the golden era of the Edo age, kamiko also became fashionable in the upper echelons of society. Members of the nobility ordered paper clothing from special workshops, but there was little left of their original, simple character by this time. The products were embroidered, printed, painted and embossed with magnificent, costly patterns and lined with the finest silks. This accorded them the status of show objects. The most precious accessories were made: gloves, scarves, big and small bags and hats. Jackets and coats were also made. Because of their lightness and resistance to water, they were often used when travelling. In the middle of the 19th century, members of the nobility stopped wearing kamiko and it was only farmers and monks who kept the tradition alive. When industrialisation began, paper clothing's importance was almost entirely lost in Japan. However, as a relic of the myth which surrounded the history of kamiko, one community of monks has kept this 1000-year old tradition alive: in the Todaji temple in Nara, during a two-week meditation cycle, young priests continue to make their own choir shirts from handmade white paper. Wearing nothing but these shirts, they retreat into nature for two months to obtain inner purity. This retreat takes place between January and March, the coldest months in Japan. On the last night, the monks return to the temple for a great feast. There is then a ceremony which marks the end of the meditation phase. Torches are lit for the ancestors

and a huge bonfire is made up on which the paper clothing is burnt. They are already bare at the knees and threadbare from the wind and weather. In this way, the monks take their leave from everything that they recognised as worthy of change during their period of inner reflection and which they could confide only to their kamiko. Cleansed by the power of the fire, they begin the new year.

Detail of an extravagantly painted and lined kamiko gilet from the beginning of the Meiji age (AD1868–1912).

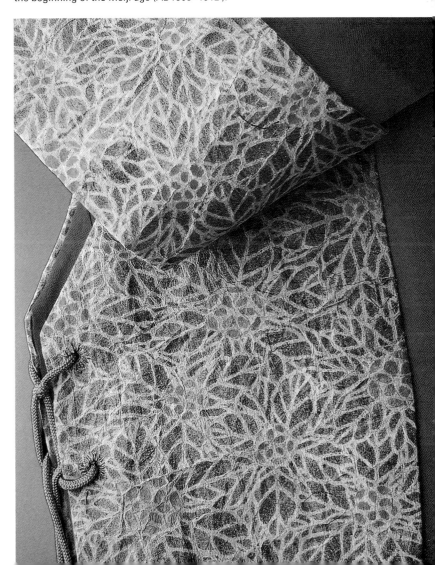

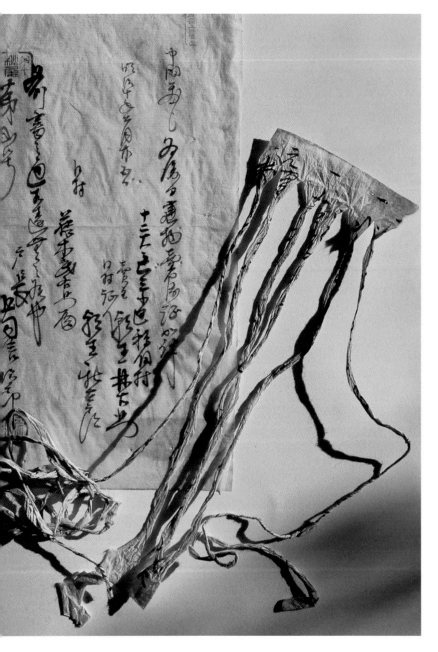

Shifu

Shi means paper, *fu* means cloth or weave. Shifu is therefore 'woven paper cloth'. This translation should be taken literally as shifu is material made from woven Japanese paper.

Using a sophisticated fold and cut system, a piece of paper is cut into endless strips which are then, either by hand or on a spinning wheel, twisted into extremely compact fibres which are then woven.[1] This brilliant, yet incredibly time-consuming process was practised for over three hundred years in ancient Japan.

Similar to kamiko, shifu was also born out of necessity. In time, the process became independent and was continually improved. Any shifu clothing from the Edo age which still remains shows evidence of a very highly developed, intellectual culture and perfect craftsmanship. Some technical skills cannot be completely reconstructed today. The paper structures were popular at all levels of Japanese society and thanks to its charisma, shifu has now become a cult object.

▲ Originally, valuable pages from old accounts books (*fukocho*) were used to make shifu because of the lack of other materials.

▶ The most precious shifu weaves start out as fine strips of high quality paper.

1 The technical details of this process are described in chapter 2, pages 50-57 .

Cultural and historical development

The earliest reference we have to shifu being made is in 1638, but its roots probably go back to the early 16th century. According to Japanese legend, a spy who had to cross enemy soil to deliver an extremely important message, is said to be the founder of shifu. The deadly secret message was written on washi, the handmade Japanese paper. The spy would have been killed, if the message had fallen into the hands of the enemy. He came up with a cunning idea to get the message through the enemy camp without being recognised: he cut the paper into strips as wide as a line of writing, twisted these into fibres and wove his clothing from this so that he could cross the foreign territory unrecognised. When he reached his client, he took the structure apart, leaving the individual sections; untwisted the fibres and was left with long strips on which the unscathed message could be read. The ruler was so fascinated by his subject's precision that he demanded that these paper textiles be produced and he called them shifu (woven paper cloth).

It is very likely that this poetic story is not entirely based on reality, but there is an element of truth as the Japanese saw shifu as a carrier of important messages for a long time.

Farmers and fishermen
Similarly to kamiko, paper weaving first started in the rural populations. Because farmers and fishermen were short of everything, they were always looking for substitutes so that they could make essential items themselves for their own use. It is very likely that the new process developed as a direct consequence of wearing kamiko. People used individually cut strips of the specially treated paper to tie clothing, lace sandals or to make bag straps. This is how they found out that the strips were more stable when they were twisted. From being used simply as loose ties, weaving the fibres into compact paper structures became standard practice.

As even paper was a valuable consumer item for the impoverished sections of the population, the people tried to find ways in which they could reuse the used paper. So pages of old accounts books and tables (*fukocho*) were used as the raw material for making shifu. It soon became clear that this emergency solution had huge advantages. Only high-quality, insect-resistant paper was allowed to be used for these important documents to ensure they lasted a long time. The quality of the material further improved after lengthy storage. The *fukocho* pages therefore had the ideal qualities for weaving. Because of the high quality of the finished materials this method remained in use for a long time. Some examples of can still be found in perfect condition

Detail of a Samurai garment made from shifu and printed with exquisite patterns.

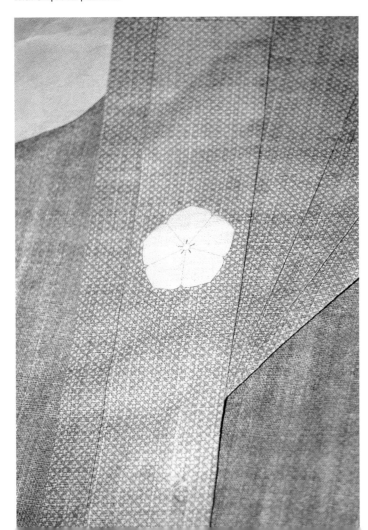

today. They can be recognised by a characteristic typical of this process: the writing in ink from the original pages is still partially visible through the twisted fibres, giving an interesting speckled pattern. The secret of the text is what gives them their special appeal. Although the text is somehow present, its content can no longer be deciphered.

Over the course of history, the making of shifu continue to be perfected. At the beginning, only individual, short paper strips were twisted by hand. Together with the weavers, the wives of farmers and fishermen, however, developed a way of obtaining a regular yarn from one sheet of paper using a spinning wheel. This was very costly, and there was often not enough time, so these structures were rather coarse – the width of the strips was between one and four centimetres. Mainly working clothes, underwear or textiles for the home were made from these.

Samurai

The Samurai, the respected class of Japanese warriors, took the shifu process over from the farmers for their own textile products and gave it a completely new meaning. Whilst the technique was a means to an end for the poorer members of the population, it then became an almost ritual act which was continually refined. Towards the end of the Edo age, the Samurai were weaving strips which were usually not wider than 2mm (1/12in.). From a 38 x 53cm (14.8 x 20.7in.) sheet of paper, they obtained yarns of over 100m (108 yards) in length. When these fine threads were woven, they acquired new material qualities and the shine and suppleness of the material covered the body in a luxurious fashion which could only be compared to finest silk. Because it took months to make a single shifu kimono, these items of clothing were extremely precious and stood for simple, unostentatious luxury.

The Samurai families were not dependent (as the farmers and fishermen were) on using the most rational method to be able to survive, as, at this time, they were very high up the social pecking order. They did not have financial difficulties or a lack of time to devote to this costly activity. After hundreds of years of war and civil hostilities, there was a long period of peace in the Edo age. The Samurai therefore did not have much to do in fighting terms. They attended to culture and produced valuable craft products as status symbols. They debated Buddhist teachings so that they could prove their position as intellectual leaders too. These priorities fitted in well with the complicated production process of shifu. As paper is seen as pure and godly in Buddhist teachings, special powers were assigned to the garments made from these. It was thought that if one covered oneself with a good, godly spirit it would have an influence on the inner being of the wearer via the clothing. The fine shifu garments were therefore almost exclusively used for ceremonial purposes.

The shoulder dress (*kami-shimo*) made from finest shifu soon became a permanent part of the traditional ceremonial garb of the Samurai. The material was printed with subtle patterns or, over months, religious texts were inscribed on the paper

Stencilled shifu weave from 100% paper

which would then be made into shifu, with the aim of reinforcing its mystical character. For the few hostilities which took place in the Edo age, the Samurai are said to have sewn their uniforms from shifu. They achieved incredible victories when wearing them and this was attributed to the protective powers of the material. It is said that the swords of their opponents became blunt so quickly from rubbing against the paper, that they were able to win every battle.

Members of the nobility
The strong, mystical significance of shifu decreased when members of nobility and rulers of the King's court discovered the advantages of the material. Around 1700 clothes made from the finest paper weaves became fashionable in noble circles and even took over from silk as the most popular and expensive material for thirty years. According to a Japanese saying from that time, '*He who has worn enough silk, changes to paper*'. The basic idea which suggested that simple shifu spoke for itself through its quality and the costly production process alone was quickly dispelled. Within a short time, the shifu materials of members of the nobility were, like kamiko, expensively decorated, lined, embroidered or stencilled.

The upper echelons of society used the paper weaves mainly to make kimonos and jackets (*haori*) or in the home for table linen, curtains and coverings. Even wide sashes (*obi*), which were tied around the kimono, and scarves, bags, tobacco cases and book bindings were also popular. Because of their status as a precious object and their ideal values, shifu weaves were also popular presents for high dignitaries in noble circles.

Farmers and Samurai made paper weaves for their own usage. When members of the nobility discovered the material for themselves, large workshops sprung up where, using a process based on the division of labour, the increasing demand was met. In the second half of the Edo age, shifu was being produced in nearly all the larger papermaking

The few remaining shifu weavers in Japan are divided between those who use the old technique traditionally and those who experiment with it.

villages in Japan. The most important production centre was Shiroishi. This town, situated at the foot of the holy mountain Zao is 300km north of Tokyo and 45km south of Sendai, the capital of this region. It is still famous today for making the highest quality paper (*zaogami*) which has been in demand for centuries in Japan and now even worldwide. Many suspect that paper weaving began in this region. For the whole of the Edo age, the valuable *shiroishi-shifu* made from *kozo fibres*, known as *torafu* there, was produced in large quantities. These were then delivered to members of the nobility and in 1719, the town was awarded a prize for this. Even today, there is still a small group of craftspeople in the surrounding area of Shiroishi who keep the shifu tradition alive.

Shifu material can be recognised by its typical burls which run through the entire material in a rhythmic pattern.

Shifu in modern Japan

Amongst the nobility, the fashion of wearing shifu clothing disappeared towards the end of the Edo period. Only farmers, fishermen and poorer Samurai families continued to use these weaves because they lacked other materials. In the Meiji age (1868-1912), when Japan opened to the outside world again following a long period of political and cultural isolation, the West began to have an increasing influence on cultural events. At the beginning there was great enthusiasm for all imported things and for several decades, this pushed traditional Japanese cultural goods into the background. At the beginning of industrialisation, attempts were made to produce shifu using machines. However, it soon became clear that this idea was not very profitable. From about 1910, the cheap import of cotton and the increasing importance of synthetic fibres, meant that these were taking over from paper weaves.

In 1921, shifu production came to a standstill in the whole of Japan and the paper required to make it was no longer produced. But in the 1940s, shifu and kamiko experienced a revival which still continues in a modest fashion today. Once the initial euphoria over imports had subsided and Western products and ways of thinking were put into perspective, a counter trend was created. A 'movement of the national consciousness' was born. Its aim was to preserve Japanese traditions and crafts which had been handed down from generation to generation. In order not to lose their own identity, craft institutes were founded (amongst many other measures) where old, traditional crafts were to be revived and scientifically documented. With the financial support of Kojuro Katakura (Lord of Shiroishi) and Masamume Date (Lord of Sendai) an organisation was founded. It began to produce shifu, kamiko and the other special papers required to make these, and thus remind the people of Japanese traditions.

In 1955, the government awarded shifu and kamiko with the prestigious title of 'Japanese cultural heritage'. From this moment on, their preservation was supported by the state. After this short boom, demand fell. Craftspeople were quickly reduced to a small group of idealists who continue to keep the tradition alive today no matter what the current trend. One of them, Sadako Sakurai, features in the gallery section of this book.

Today in Japan, there are about ten to twenty, elderly shifu weavers who work with the old technique in very different ways, some keeping it

The weft and the warp of this stencilled material which is over 60 years old is made from paper (*morojifu*).

If the Shifu weft thread is alternated with other materials, as in this structure by Sanja Hanemann (Austria), then an interesting ribbed structure is produced (*kobai-ori-shifu*).

Qualities and types of shifu materials

In the golden age of Japanese paper weaving production, it was thought that a spirit inhabited the shifu garments and that it was continually swapping places with the wearer of the item. Indeed, once you have discovered the outstanding wearable qualities of this material, the idea no longer seems so absurd. Shifu weaves react more intensively than other materials to body heat and evaporation. They can become bigger and smaller and adapt to suit body shape. In the summer they are light, like linen, and can absorb a good deal of moisture, soaking up sweat without sticking. Closely-woven materials are very warm in winter because of the isolating effect of the paper. They cannot become electrically-charged and, depending on which fibres are used for producing the paper, they can even have an antiseptic effect.

Finely woven shifu materials flatter the body and have a particularly nice drape because the precious shine and softness of Japanese paper is retained even when twisted and woven. The weaves therefore always have a matt, shimmering effect and an interesting, grainy surface quality. Their texture can best be compared to raw silk. The irregular weaving structure is also characteristic of these materials. When the sheet of paper is cut up, there is a surplus of material on the turning points along the external sides. After twisting, the thread is thicker in these areas. In the finished yarn, they can be seen as regularly-arranged burls and, when woven, these run through the entire material adding rhythmic accents to it. This brings the surface to life and structures it from the inside out.

By twist contracting the strips regularly, the paper becomes much more stable and it produces unexpectedly strong yarns. But because shifu yarns are extremely non-elastic and tear quite easily, weavers need a good deal of practice and sensitivity. The correct rhythm, careful setting up and the ideal humidity are very important. Some paper weaves are given further special finishes when they are taken off the loom so that they lose their stiffness. After they have been worn a few times, the material usually

strictly traditional, others preferring to experiment. Although they are few, their work has had a subtle effect: today, there are people from the Philippines to America and from Nepal to Europe who are fascinated by the idea of shifu and have decided to continue to spin the fibres in their own way.

becomes softer anyway. They can be washed and cleaned without any problems and they are incredibly stable. The durability of paper is evident from historical shifu-blend weaves. After 200 to 300 years, when the silk or cotton warp of old materials has already rotted or been eaten away, the paper weft is still intact. Natural substances in paper prevent insects and bacteria from attacking it and make the material surprisingly durable.

In ancient Japan, pure paper weaves were made in which both the warp and the weft were made from paper fibres. They were called *morojifu* (or *moroshifu*). They were very expensive to make and were therefore seen as a precious commodity. Weavers needed a lot of experience and sensitivity to be able to make a *morojifu*.

More often, different types of blended weaves were used, where paper threads were always inserted. They had different names according to the combination of materials used: *kinujifu* (warp: silk, weft: paper) and *asajifu* (warp: linen, ramie, hemp or other raffia fibres, weft: paper).

Poorer parts of the population wore mainly coarse, simple weaves (*hira-ori*), whilst the elite wore simple shifu materials with complicated patterns. Materials in which the shifu thread had been alternated with another material in the weft were also popular (*kobai-ori-shifu*). For example, you would weave a few wefts of the silk warp material rhythmically into the silk warp and then you would weave somewhat thicker paper yarns in as an effect-yarn to produce a ribbed surface structure.

Crepe-like shifu weaves which were mainly used for summer kimonos have an unusual quality. To make these *chirimenjifus,* slightly overspun paper threads were woven. By using a clever combination of materials and special finishings, the threads gather together and give an interesting, bark-like, pleated structure. There are only a few remaining examples of these left, but their special beauty is particularly impressive.

One particular Japanese way of using paper to make textiles gained popularity in Western culture quite early on. Since the Edo age, it was usual in Japan to weave gold or silver-plated paper lamella (*hirahaku*) and spun paper threads, coated in metal (*yorihaku*) into magnificent materials. These precious materials with special effects were exported to Europe. There they were made into the most expensive brocades under the name of *Japanese gold*. It was well known that paper could be thinly coated in precious metals to produce textiles. It had been going on for centuries in the West, but it was not until the end of the 19th century that people began to twist the material into yarns and view it as a textile material in its own right.

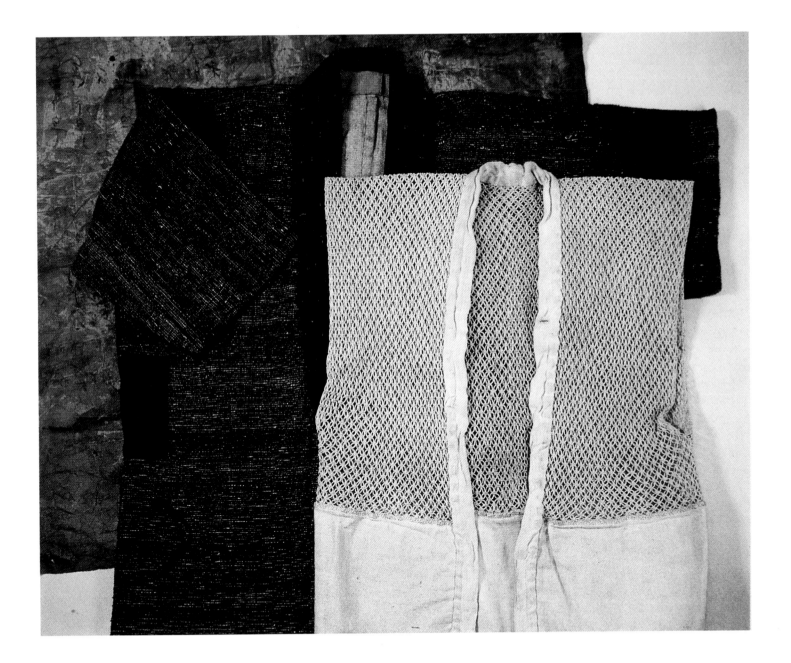

▲ Coarsely woven farmer's workcoat made from accounts books
and a warrior's buttoned undershirt made from the finest shifu.

◄ Weave made from gold-covered paper strips (*hirahaku*) from the
end of the Edo age

1.2 Paper textiles in Western culture

In Europe and America, using paper for different aspects of daily life is not so strongly embedded as it is in Japan. Despite this, textiles made from paper have a history which dates back from than 100 years. It was much more important in the first half of the 20th century than we are aware of today.

Paper linen

In the middle of the 19th century in the West, because of a lack of other available raw materials, people began to use paper as a substitute material for making textiles. In the same way as sheets of waterproofed paper were used to make the Japanese kamiko initially, they used them to make clothes and accessories. However, Western paper had a completely different character to Japanese because it was made from wooden fibres and, at this time, was already industrially manufactured. The material was therefore stiffer, but also more brittle than that of Japanese paper clothes, and its character was anonymous. It was rare that entire garments were made from it. Instead, it was used primarily to make shirt collars, underwear, ties, cuffs and bed linen. The products made from this sturdy, waterproofed paper had the advantage that they could be produced industrially very cheaply and could simply be disposed of after use. Paper linen was originally invented in the US, where this new form of consumerism soon enjoyed great success. The *linen reform* soon spread to Europe. Towards the end of the 19th century, men's shirts made from simple cotton tricot became fashionable. Paper collars, cuffs, and undershirts (known as *serviteurs*, they covered the front part of the chest) were made separately and could simply be buttoned on. The advantage of this was that individual parts could be replaced or cleaned separately and costs for a new wardrobe could be reduced. Lacy patterns or weave structures were often printed onto the paper to increase their textile-like effect.

Even the clothes the poorer people buried their corpses in were made primarily of paper at this time.

Paper collar made by the company Mey & Edlich, Leipzig.

Particularly during the war years, people could not afford to bury the dead in clothes made from fabric. Up until the 1960s, the use of paper shrouds was standard practice amongst poorer people. For ecological reasons, these shrouds have become popular again over the last few years. Aside from this, however, paper linen has almost entirely disappeared from daily use.

The paper yarn industry in wartime Europe

In addition to the paper linen industry which was, on the whole, rather insignificant, at the end of the 19th century in the West, a process was developed for working with paper in the form of yarns. In wartime and between the wars, when other raw materials were in short supply, this branch of the industry grew and became important economically.

In contrast to Japan, the manufacture of paper yarns in the West was an industrial process from the very start. A sheet of paper, limited in size, is used as the source material for making shifu threads. In Western paper yarn production on the other hand, the multi-filament rolls which come out of the paper machine are processed. Special cutting devices divide these rolls of paper into thin strips which are fed straight onto a spinning machine where they are spun into a yarn and wound on a reel. The thread does not have the burl structure seen with shifu. It has the regular, unitary character of industrially-made products. European paper yarns are stiffer and harder to the touch giving them a certain expressiveness and a particular aesthetic quality. In contrast to Japan, spiritual or religious content has never played a role in paper production in the West.

The first coarse paper strings were used as packaging or binding material. Sheafs of grain were tied together with the strings shown in the picture.

The beginning

There is little known about the origins of Western paper yarn manufacture. It is assumed that it began in the US. But the idea had already been brought to Europe by Dr Alexander Mitscherlich, from Germany, in 1890. It was taken up straight away and further developed. Germany soon became the worldwide leader in this young branch of the industry.

The German textile industry was a very successful and growing branch of the economy at the turn of century. Whatever was needed in Germany, could be made in Germany, and export continued to rise. In spite of this, Germany was confronted with a problem: because of its lack of raw materials within Germany, it was almost entirely dependent on imports from abroad. Large quantities of cotton, jute, hemp and linen had to be imported from various countries throughout the world. In Germany therefore, people were more concerned than in other countries with developing an alternative raw material, a surrogate, which could be found in Germany. Given these circumstances, paper yarn seemed very promising.

Europeans were well aware at this time that the Japanese had a very high paper culture and that they were able to twist this material into fine fibres and make soft weaves from them. The Swedish nature reseracher, C.D. Thunberg, wrote about the Japanese shifu materials as early as 1775.

The weaves also received a lot of attention at the World Exhibition in Vienna in 1873. This was the first time they had been seen in Europe and they led the German University professor, Dr. J. J. Rein, to take a research trip to Japan. In his comprehensive work, *The Industries of Japan* from 1889, he gives a detailed description of the then-flourishing kamiko and shifu culture in Japan. On the basis of this report, two delegates from the German embassy visited Shiroishi (the centre of Japanese shifu production) in 1892 to get inspiration for the German textile industry and to drive the replacement material onwards.

In 1913, there were already eight large manufacturers who specialised in paper yarn production and making. They achieved a high turnover by producing various technical weaves such as bags, straps, saddles or packaging strings. As it was light and dust-resistant, the material was ideal for these purposes. When the First World War broke out in 1914, and raw materials for daily textiles became even more scarce, people were forced to refine the material so that they could use it for other products also.

The First World War

Because of the devastating political and economic situation during the First World War (1914-1918), from 1915 onwards, paper strings became increasingly important for day-to-day living in Germany.

Enemy nations had issued Germany and its allies with an export ban so the import of foreign goods almost came to a standstill. Groceries and raw materials became even scarcer. The situation was particularly dramatic in the texile industry. In addition to the usual production of clothes, they had to meet the increased need for uniforms, blankets and tents on the front. However, they lacked the necessary textile fibres to make them. From 1915, the search for a possible surrogate began in earnest. A variety of native plants, such as nettle and reed fibres, were examined to determine whether they were suitable to be made into yarn.

The paper yarn industry received more attention at this time, because paper could be produced in quite large quantities within Germany. New technologies were then developed which made it be possible to produce very fine paper yarns from which textiles for clothing or the home could be made. Outer and under clothing soon followed, with padding and lining materials, nightshirts, underwear, corsets, hats, belts, braces, bags, shoes and, for the home, furniture coverings, woven seats, tapestries, blankets, carpets, table and bed linen, curtains, rugs, handtowels and much more. Military, marine and administrative uniforms were also made from this surrogate material. No less than 600 tons of paper weave were produced per month in Germany between 1915 and 1918.

Although the newly-developed paper yarn was thin, it was still not at all elastic and was hard to the touch. The quality of the product was relatively low on the whole, particularly in terms of comfort and washability. This is because its original character as a packaging and tieing material never entirely

Child's vest from a paper weave, Germany, 1916/17 (with thanks to the Germany History Museum in Berlin).

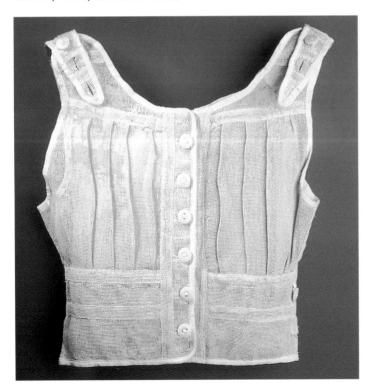

disappeared. The products were therefore rather unpopular with people. The government nonetheless tried to increase the value of the material using propaganda and promoting it as proof of the efficiency of German industry. In a journal from 1918, it was written,

'we must sing a song of thanks to paper in its various forms, because it has helped us to create virtue from need. An entire fashion industry has built up overnight from using this paper for clothing. With this new material, we can achieve effects which completely belie its source [...] we are packed in paper and sent as a surprise parcel into Spring. These parcels, when unpacked, provide dazzling evidence of German diligence and industry' (quotation from: Wachtel, Joachim (Hg.): *'À la mode. 600 years of European fashion in contemporary documents'*. Munich, 1963, pp 245ff).

Despite these attempts, paper's image scarcely improved and once the war was over, it disappeared quite quickly from day-to-day life. Production dropped drastically between the two world wars. Only a few factories continued to make the material and thanks to their quietly pioneering work, contributed to the ongoing improvement of the quality of the yarn.

Before the Second World War broke out, the process was well-known throughout Europe and some other countries had also begun to produce it.

The Second World War
Over the years, Germany's economic dependence on other countries had again increased. When the Second World War (1939-1945) broke out, the situation seemed as hopeless as it had done during the First World War. International border blockades made it largely impossible to import raw materials and forced Germany and its allies into economic isolation.

In addition to these circumstances beyond their control, the NS (National Socialist) regime also pursued a fatal, ideological concept. Its aim was *autocracy*, i.e. economic and cultural independence from abroad. Foreign influences which were seen as inferior were to be virtually eliminated in order to create a pure 'aryan industry' for the German people. Step by step the German economy was to be restructured so that it could provide for itself.

Yet again, people had to pinch and scrape to make textiles. The search for surrogates did, in fact, force the development of artifical silk as well as the first synthetic materials, such as nylon and perlon. Again, people relied on naturally-occurring replacements from the native plant kingdom and fostered the manufacture of paper yarn.

In the meantime, the material for paper yarns had developed further technically. It could now be made watertight and its somewhat dull appearance had been improved with various finishing processes. The quality of the yarn was therefore much higher than it had been during the First World War. The rather stiff weaves were being used more specifically for interior furnishings or accessories for which the material was better suited. Since about 1930, it had also been possible to manufacture knitwear from paper strings on machines. *Pextil jersey* was made on circular knitting machines. The resulting stitching was softer and stretchier than the stiff weaves and was therefore more pleasant to wear next to the skin. Mainly hosiery, nightshirts, vests and drapes were made from it.

Despite these technical improvements the material held on to its bad reputation from the First World War. The government kept on trying to convince the public of the transformation of the paper yarn '*from substitute material to plain material*'. They tried to change its name from 'paper yarn' to 'cellulose yarn' to get rid of the negative associations. For the duration of the war and post-war period its image changed very little.

Although the nature of the substitute material underwent strong growth, the NS regime never achieved the autocracy it wanted. Great shortages of material and devastating poverty in almost all areas

Advertising poster for a paper yarn exhibition in Vienna, 1918 (with thanks to the Rhine Industrial Museum, paper mill at Alte Dombach)

of life affected the war economy making private consumption even more limited than in the First World War.

Even after the war ended, the situation did not improve until 1947/48 when reconstruction began which then led to the economic upturn and boom. When imported fibres were finally available again, people quickly turned their backs on the inferior substitute materials and disposed of, or burnt, the old textiles to avoid wartime memories. For this reason, exhibits of paper string products from this period, such as written documents about the manufacture and making of the material, are a rarity today.

Various bags made from paper yarn, Germany, 1930 to 1965
(with thanks to the Rhine Industrial Museum, paper mill at Alte
Dombach)

Doll's pram made from paper yarn, about 1950
(with the kind permission of the Rhine Industrial
Museum, paper mill at Alte Dombach)

Finland

The situation in Finland was somewhat different. A wealth of historical paper weaves remain today. This is because of the special political position this country holds. Finland is a young nation and was heteronomous between Sweden and Russia alternately for a long time. However, in 1948, Russia which governed Finland up until this time, recognised it as an independent state. However, they had to pay high reparation costs to Russia until the mid-1950s. Large amounts of materials and raw goods had to be handed over to the former ruling country. Amongst these was a large proportion of the textile fibres available in Finland. So whilst the rest of Europe enjoyed economic recovery and forgot about the wartime substitute materials, Finland suffered from poverty and a lack of raw materials until the end of the 1950s. Finnish textile makers had to use the substitute materials for far longer than those in other countries. Huge quantities of cheap paper yarn were also made as the material was seen as inferior and therefore did not have to be handed over.

In Germany, the deprivation of the war did not allow for creativity, instead it gave rise to specific industrial manufacture. Paper strings in the post-war years in Finland were made to a high technical and creative level. Instead of using cotton, linen or silk, the textile makers expressed their ideas in the substitute material. The general euphoria which surrounded their independence released a great deal of creative and artistic potential. The textile sector made a great contribution to establishing an independent, Finnish identity.

Whilst articles were being produced en masse in German factories, and their designers remained unknown; in Finland, well-known designers were producing designs and getting them made either in their own workshops or in large factories. Because of a lack of alternatives, the brightly-coloured paper strings were often combined with other, unusual materials such as plastic strings, acrylic glass rods, birch bark, synthetic strips of wood, raffia or similar. The original samples inspired the production of lamps, bags or furniture coverings.

The most famous of these textile creators was Dora Jung. In 1932, she set up her own workshop. She produced mainly home furnishing materials and art textiles made of linen. She received many medals at world exhibitions for her subtle tapestries.

Because linen was in short supply in the wartime and in the post-war period, she also reverted to using paper yarn which she used to make into costly, patterned materials. The preserved damask weaves demonstrate her enormous ability and her extremely polished technique, right down to the last detail. It was only when the economic situation in Finland improved towards the end of the 1950s that paper strings began to disappear there too. The complete paper yarn collection by Dora Jung as well as many other paper samples by other textile artists from this period, such as Aino Keinanen-Baeckmann, Eva Taimi and Margareta Ahlstedt-Willandt, have been preserved and are kept in the Helsinki Design Museum. Because of its history, it is not surprising that the paper string boom, which is currently spreading throughout the whole of Europe, began in Finland in the 1980s.

Shoes made from plaited paper strings, Finland, about 1940 (thanks to the Workers' Museum in Alt Amuri in Tampere)

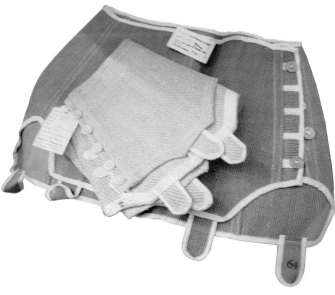

▲ Close-up of a sofa covering made from 100% paper, printed
  with a pattern by Eva Taimi (Finland), about 1950 (with thanks
  to the Helsinki Design Museum)
◄ Corsets made from 100% paper, Finland, about 1940
  (with thanks to the Helsinki Design Museum)

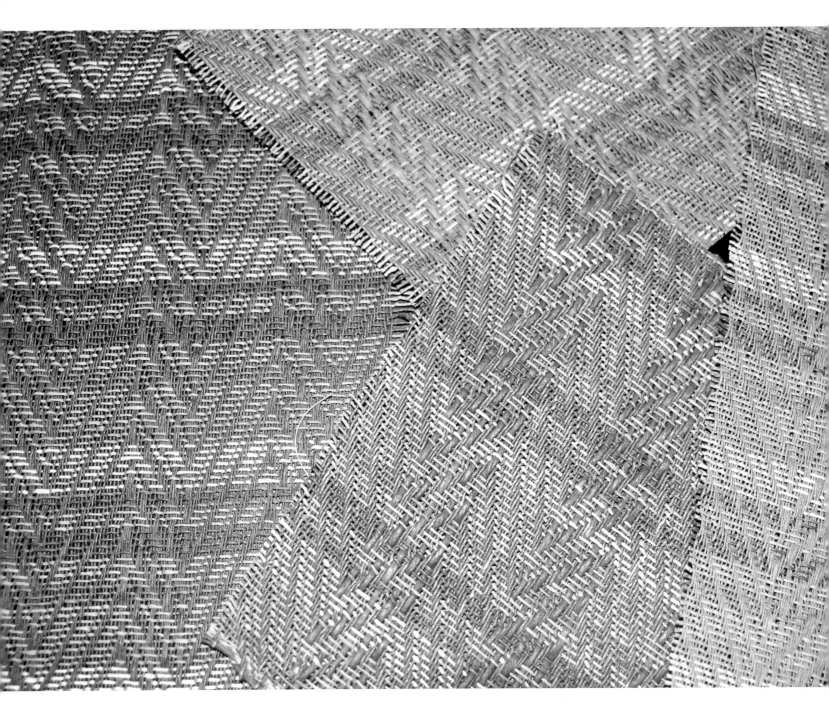

Paper weave samples by Margareta Ahlstedt-Willandt (Finland), about
1950 (with thanks to the Helsinki Design Museum)

Damask weaves by Dora Jung (Finland), 100% paper, about 1950
(with thanks to the Helsinki Design Museum)

The Lloyd Loom company

L*loyd Loom* occupies a special position in paper string history. Since the beginning of the 20th century, this American company has been producing furniture from wire-strengthened paper string, and is still doing so successfully today. This company provides therefore a bridging point from the history of paper textiles to current developments in this area.

The CC *White Manufacturers Company* originally produced children's prams made from woven cane. As was the case with all wickerwork, these could only be made by hand as the brittle natural material could only be made into a three-dimensional form using an organic process, together with the wooden frame. They were therefore made very lavishly and this was reflected in the price. During the industrialisation period, the competition, who were working with different materials, became so strong at the beginning of the 20th century that the company almost had to close down.

The new director of the company, Marshall Burns Lloyd (1858-1927), a legendary sort of inventer, saved the company from ruin in 1917 when he introduced a reform which made the manufacturing process considerably easier. Instead of the brittle, unmanageable natural material, they began to use relatively thick, strong paper strings for the weave instead. A wire core was fixed on the inside to increase its stability. In order to work with the material, Marshall Burns Lloyd developed a special web chair (loom), known as the Lloyd Loom, which made it possible to industrially produce lengths of weave which were very similar to woven rattan. The weave only came in the form of piece goods off the loom and had to then be further cut to size, stretched across a prefabricated, wooden-strutted frame and bent into shape. The edges were nailed on and the crossover points covered with decorative selvedges.

This new process had many advantages: production became much cheaper, because many parts of the production process could be automated. To make a pram from paper string, only one fourteenth of the original time was needed. The source material was also much cheaper as it did not have to be imported from far-away countries as cane did, but could be obtained from domestic paper spinning. It was also no longer necessary to take the natural length of the plant material into account. The paper string could be produced as a multi-filamented thread so that strips of any width and length could be made. The new weave was also lighter, more pliable and could be dyed more easily than the original material.

Marshall Burns Lloyd was, of course, aware that the people did not like paper yarn as they associated it with wartime necessity. He therefore sold it as a 'new sort of fibre' and did not mention that paper was the source material. As only a few people knew the truth, the once inferior substitute material successfully acquired a stylish image.
All of these factors led to a huge increase in sales over a few years. The company, renamed Lloyd Loom, was soon one of the most important producers of prams in the US. They then began to make furniture from paper string. The success of this soon overshadowed their prams. Various prototypes of simple, elegantly proportioned chairs and tables were put on the market and these soon became design classics. They were seen as being long-lasting and comfortable and they did not creak. Thanks to the layer of varnish, they were also water-resistant and could therefore be used both outside and inside. The Loom furniture was also popular for equipping hotels, patios, boxes in stadia, deep sea boats and airships . They were particularly suitable for the latter because they were very light.

The demand for Loom products soon increased in Europe too. In 1922, the English furniture company, *Lusty & Sons,* bought the patents from Lloyd Loom and took up production and soon achieved a similar level of success. However, after several brilliant years, demand fell sharply throughout the world in the 1950s. New materials like chrome, steel and cheap synthetic materials were seen as more modern and were accompanied by new designs. Shortly after the closure of the parent operation in the US, the English company also had to stop production at the end of the 1950s.

As people became more environmentally-conscious, natural materials like wood and basket-making materials regained currency. People rediscovered the old Loom furniture. As sought-after collectors' pieces they fetched very high prices at auction, further adding to their once stylish image. It was not difficult for the former manufacturer to

capitalise on this myth. In the middle of the 1980s, the original firm in the US started to produce the old classics again. In Europe, the *Accente* company from Germany took the manufacturing rights over from *Lusty & Sons* in England. Side by side with the old models, new designs were successfully sold throughout the whole world. Even today, only experts know that the originial Lloyd Loom furniture was made from paper strings.

Classic Lloyd Loom furniture

2 Different material qualities

▲ Different stages of making shifu
◀ European paper strings

There are two, very distinct main groups of paper yarn. There is the paper yarn which can be made by hand according to the old Japanese shifu tradition and which produces a very individual, soft fibre with burls, which are typical of this technique. Then there are the many varieties of paper strings from wooden fibres which are machine-made. These have completely different characteristics: they are stiffer, more robust and have the anonymous character of the mass-produced product. Both materials have their own special charm and make inimitable effects possible. To make things simpler, all industrially-produced yarns will be referred to as paper yarns and all the fibres which are made by hand will be referred to as shifu yarns or fibres (although we know that these are paper yarns too).

◄ The choice of paper is extremely important for the quality of shifu-Garns.
▼ Various stages of making *kozo*: the bark of the paper mulberry bush is the usual raw material for Japanese paper.

Handmade shifu yarn

Making shifu yarns is a long-winded process. Before the fibres can even be formed, the right paper has to be created. This first step alone, which covers planting the bushes to drying the paper, is very time-consuming. Even the skilled method through which an 'endless' (multi-filament) yarn is obtained from one piece of paper is both time- and work-intensive. However, the fact that in the production process, every step can be influenced, and that you are aiding the design process every step of the way, makes this material valuable. Even if the process seems unnecessarily awkward, the charm of the material more than makes up for this.

Choosing the paper

A good shifu thread begins with good paper. Because the character and the qualities of the basic material are directly reflected in the thread, the choice makes all the difference. When choosing the paper, you should always think what you want to use the yarn for later. For classical shifu production according to the Japanese model, certain qualities with special characteristics have many advantages. They emerged from a long tradition and have lasted centuries. Even if you want to experiment freely with unusual papers, it is still important to understand these foundations first so that you are able to produce something really new from them.

Japanese papers

The quality of a shifu thread, made from handmade paper, is, as a rule much higher than that of industrially-produced paper. Unfortunately, over the last few decades, the craft of traditional papermaking in Japan has declined drastically. Today, many Japanese papers are already completely or partially industrially-produced, but they continue to be of a very high quality. Truly high-quality paper which produces a beautiful, soft shifu thread is simply not available here (in Europe). So that you can recognise them in a trade shop, it is useful to know the basic criteria which determine quality. Today, there are many sorts of decorative paper on the market which have a very decorative effect achieved by inserting various plants or an irregular arrangement of fibre bundles. They are often sold as 'Japanese paper'. Even if it can be said that these papers are of Japanese

origin, the quality of these papers is unsuitable for classic shifu yarn production. The simplest, most delicately-fibred, regular paper available is what you need.

When you buy Japanese paper, you can tell a lot from the plant from which the base fibres of the sheet were obtained. In good specialist paper shops, this should be displayed or if not, you should be able to ask the assistant. The *kozo* bush (paper mulberry bush), often just called *kozo*, provides about 80% of the raw material for making Japanese paper. The very long fibres, known as *yuu*, are obtained from the inside bark. The papers are very stable and have a lot of body.

The bush-like plant, *mitsumata* (*Edgeworthia papyrifera*) belongs to the family of daphne plants and provides the fibres for very soft, shiny paper. These are primarily used for prints and woodcuts, but also for paper money, amongst other things.

Gampi (*Diplomorpha sikokiana*) is also a member of the daphne plant family. It is seen as the most precious of the plants providing the raw materials. The papers are transparent and similar to silk, but are also very stable and are easy to write on. The bushes, which are virtually impossible to cultivate, grow in the wild in the mountains and can only be pulled every five to six years. All three of these fibre types, which are sometimes blended, are very long-lasting and insect-resistant. The papers made from them are therefore much softer, but also tougher than our Western paper which is made from broken-down wood. In principle, all the papers made from these three plant fibres are suitable for making shifu yarn, but the most used and also easiest to get hold of, is *kozo*. The *kozo* plant is reaped in Autumn, then dried and often stored for several years. Before it is used, the branches are left in water over night, the outside bark is peeled off and the bast is boiled for about three hours in a potash or washing soda solution until the fibres open up. In soft, cold water the bast is rinsed until the residue has gone and it has returned to its neutral pH. This usually takes place in the winter months in the mountain regions, where the water has the ideal properties to facilitate this. The next step is to beat the fibres with wooden hammers or special pressing machines, so that it disentegrates into individual parts. Finally, water is added to these, so that it can be made into a pulp. Because the fibres are not cut using this method, but keep their length, even thin sheets are proportionally tough and stable. In the industrial opening-up process, the chemicals used are often more aggressive and they bleach the paper and can no longer be rinsed off. The fibres are also broken up in the *hollander*, for example, where they are cut up into short pieces. Good handmade paper should, however, have a neutral pH and the longest fibres possible. Sometimes sheets are sold as *kozo* paper even if they only really contain a small part of this base fibre. However, you should usually be able to get information in specialist shops. You should be careful when buying 'Japanese papers' if the base fibres are not given or if you cannot ask what they are. The quality of a paper which is suitable for making shifu, is not just determined by the fibres used, but also by the way it is made. It is only possible to recognise the important characteristics of handmade paper if you have a rough idea about the traditional Japanese papermaking method.

The Japanese used a different papermaking sieve to the one used in the West: the *suketa* is usually a right-angled sieve made from fine, bamboo shoots (*su*) laid in parallel lengthways. The shoots are fixed together vertically with fine silk or nylon threads. This *rolling sieve* lies freely within a wooden frame (*keta*) which has a basic structure and a lid. This is fixed onto one of the lengths with hinges. With the Western sieve, the lid is taken off completely once the fibre paste has been removed and the sheet of paper is couched from the stiff sieve. In Japan on the other hand, the sieves, which are often very large, are hung on a sprung device above the tub. After the paper has been made, only the lid of the frame is opened and the rolling sieve with the non-woven fleece is removed from the frame. The sheet is laid onto a pile in one fluid movement and thus released from the rolling sieve.

The most widely used papermaking method is the *nagashi-zuki*, which translates as 'multi-layer process'. In the first step, the sieve is dipped into the pulp on the long side and fibre mass is absorbed. The sieve is then swung back and forth and the excess water is dispensed of with a backwards movement. Then the sieve is dipped again, picks up some pulp and the sieve moves. When there is almost no fluid left, the sieve is dipped into the pulp again

immediately and the next layer of fibre is built according to the same principle. This process is repeated until the desired level of thickness is reached. Finally, pulp is taken up again, the sieve is swung back and forth and the superfluous fluid disposed of over the front edge of the sieve. The thickness of the sheet in the Western method is determined only by the amount and consistency of the pulp when the sieve is dipped in one time. A Japanese papermaker, however, can produce different thicknesses of paper from the same fibrous mass. Each papermaker lends his papers an individual character. Depending on what the sheet is to be used for later, different methods can be used. It is only possible to repeatedly dip the sieve when making a single sheet because the Japanese pulp is enriched with *neri*. This is a slimy extract from the root of the native Japanese hibiscus plant, *tororo-aoi*. This viscous substance aids the formation of the sheet and makes the fibres move regularly around the pulp. It stops them from clumping and makes the water drain off more slowly and viscously from the sieve. This makes it possible to create several layers and thus very thin papers which are still full of body and have a luxurious shine. When the single sheets are couched on top of one another after the papermaking, because of the *neri,* it is not necessary to put inserts beween the sheets as they do not stick to one another. Even if a whole pile of several hundred sheets were slowly drained of all their water, they would still be able to be separated easily. The sheets are then spread onto a large wooden board using a wide paper brush. After their surface has been smoothed, they are left to dry and bleach in the winter sun. This is how they obtain one smooth and one slightly rougher side. Today, the sheets are often put onto heated metal plates to produce two equally-smooth sides.

To determine whether or not a Japanese paper has been handmade or not, you can check whether the marks from the rolling sieve can be seen on the sheet when you hold it up to the light. On a square piece of paper, if you can see vertical lines (from the sieve binding) and fine, horizontal hatchings (from the individual bamboo shoots), this is a good indication of a craft product. Unfortunately, this sieve imprint is sometimes very weak and not always conclusive. Mesh-like sieves may also have been used and these leave a different imprint. These papers are also suitable.

Another feature which hints at a craft product is the course of the fibres. In the swinging movements, the fibres are not randomly arranged, they have a direction. If you can see single fibres up to 2.5cm (1in.) in length on a sheet, and these nearly all run parallel to the opposite side of the sheet, it is very likely that the sheet was made by hand. The format and the edge of the sheet of paper can also tell you whether the paper was made by hand. Although there is no standard format for Japanese paper, over time, certain formats became more common, in particular sheets measuring about 94 x 63 cm, 54 x 38cm and 70 x 136cm (36 x 25in., 21 x 15in. and 27 x 53in.). The famous workshops tended to make larger formats. If the sheets have four, somewhat irregular edges (caused by the restriction of the frame of the sieve), then it is very likely that they have been made by hand. Especially if the edge of the tub is thicker down one of its long sides, then you can be sure that a rolling sieve was used because this material thickening will have left a mark on the pile of paper when the sieve is put down. The few Japanese businesses which still exist guarantee the highest possible quality. Last but not least, you can tell good paper by the relatively high price!

In Japan, shifu weavers ordered papers from their own personal papermakers. The knowledge of how to make these special papers was the preserve of only a few papermakers. *Shifugami* was traditionally made from *kozo* and was made so that all the fibres lay in parallel to the opposite side of the sieve. They were not cut in two when the yarns were formed because the strips were always cut parallel to the direction of the fibres. The finished thread was therefore very tear-proof. The low stability of the paper across the width did not compromise the quality of the yarn.

Papers which have been specially produced to make shifu are not available here in the West and, even in Japan, there are only a few places which still make them. Because the majority of Japanese papers are made using the *nagashi-zuki* method, nearly all Japanese papers which have a regular structure and weigh between 18 and 28g/m^2, are suitable for making shifu yarn. It is only possible to tell how the material will react and whether it will produce a nice, strong thread, when you have tried it!

Nepalese papers

Nepal has a long tradition of making high-quality paper. In the past, however, it was never used to make yarn. The fibres for the pulp, *lothka*, are obtained from the bark of a

Traditional Japanese paper production using a rolling sieve

| Take up pulp | Open lid | Remove sieve | Couch non-woven fleece |

native daphne bush, the Nepalese paper daphne or *Daphne papyracea* which only grows in the mountains at between 1800 and 3000m above sea level. These plants have wonderful qualities which are transferred to paper and later to the woven material. Because of their antiseptic effect, paper has always been used by the Nepalese rural population as plasters. They are also pest-resistant which makes the paper particularly long-lasting.

In many parts of Nepal, paper is made according to the so-called *pouring process*, i.e. they are not made, but poured. After drying, cleaning, boiling and beating the loktha fibres, the pulp is prepared and the exact amount required to make one sheet is emptied into a simple sieve which floats in a pond with in and outflow. If there is enough fibre pulp in the sieve, then it is carefully removed and the superfluous water is allowed to drain and it is put out in the sun to dry. If you use this method, the papers are coarser and sometimes have different levels of thickness.

They only have one layer and a natural, rustic look. As they are not couched to make several sheets, several sieves are needed which makes this pouring process rather costly. This is the original method used in China to make the first-ever paper. In Nepal, and, to some extent, in Bhutan, this process still exists today.

Although Nepalese paper has a completely different character to Japanese paper, they have very similar qualities which is why *lotktha* papers are also suitable for making shifu yarns. Twenty years ago, a Nepalese man, Deepak Shresta, began to make this paper into yarns and materials. Since he began to export his products, the use of *loktha* paper to make shifu has also become common here. This paper is usually cheaper than the Japanese paper and is easier to get hold of because the decorative, often brightly-coloured paper is currently in fashion and can be bought in many paper and alternative shops under the names 'Nepalese', 'Himalayan' or 'Loktha paper'. They are

▶ In many parts of Nepal, paper is still made using the traditional pouring process
▼ Nepalese *loktha* paper is increasingly being used to make textiles.

Imprints left by the rolling sieve and the irregular tub edge are indications that a paper has been made by hand.

available in various strengths, but a density of about 20 to 30 g/m² is ideal for making shifu yarn.

In the pouring process, the fibres do not show a particular direction, rather they have a completely random arrangement. You can see individual, dark fibre bundles with the naked eye on the yellowy-brown paper. In order to be able to work the paper into yarns, you should make sure that they have been poured as regularly as possible and that they are about the same thickness all over. Thick accumulations of fibres, wooden hinges or areas with holes are not suitable for making yarn. Apart from the above, this quality of paper can be worked without any problems. It produces natural, full-bodied threads which are full of character. This material is ideal for learning the technique because it is not very sensitive and hardly needs any

moistening. Exciting experiments are possible too, if you write on it, paint or marble it, or pre-treat the paper in similar ways. The resulting qualities of yarn are attractive in their own right and open up new possibilities for using paper in the textile sector.

Experimental paper qualities

In principle, almost any paper can be made into shifu threads. Asian paper whose fibres are obtained from their native plants is usually softer and more elastic, thereby producing workable threads. For example, the Japanese who work in the Philippines produce wonderful threads from pineapple fibre paper. There is even interesting paper from Western regions available here. It is not always possible to find out about its origins and base substances, but it is worth experimenting with.

Handmade shifu threads can also be made from industrially-produced paper. Wood fibre paper which is common over here in Europe, is made up of very short fibres which are tough and sturdy. When it is twisted therefore, it does not form closed, regular threads. It is rougher to work with, but can still give interesting effects. Experiments using transparent paper, crepe paper, packaging material, silk paper, coated materials or printed photos or written pages can also give interesting results.

There are no limits to experimenting. Even paper which seems quite unstable, can work, because twisting the paper threads makes the paper somewhat stronger. The thickness of the yarn always depends on the paper strength and the width of the cut strips. You get to know each paper quality in a new way when you to try to transform it into a yarn. But even unspun, paper can be used as an exciting weaving or braiding material. If it is only cut into strips, for example, interesting structures are produced when they are woven into a thin warp. For example, pieces of cut up picture material inserted at intervals, or strips which have been crushed by a hefty blow can achieve new sorts of effects. You should always choose your material according to what you want it to express and with its end use in mind.

Nepalese paper dyed with natural dyes.

The individual production steps

When you have decided on a particular paper quality, you can begin to make the shifu fibres.

Materials required:
- Shifu paper
- Set square, pencil
- Something to rest on when cutting
- Iron ruler
- Stanley knife or cutter
- Spray bottle

- Bamboo mat or a similar firm surface
- Screw clamp
- Dishcloth
- Plastic bowl
- Small plastic bag
- Spindle or spinning wheel

Folding

First, a whole sheet of paper is folded in a special way: once in the middle, parallel to the long side. Then, both halves are again folded to the bow which has been produced. At this point, the edges should extend about 1.5cm (3/5in.) beyond the central fold. A w-shape can be seen in the cross-section. It is important to make sure that the thread is running the correct way with Japanese papers. The three bows must always be folded parallel to the bamboo hatchings from the sieve imprint, i.e. across the imprints from the sieve binding, because the direction of the fibre runs parallel to the narrow side of the sheet. At the beginning, depending on the format, it is recommended that you use only 1/4, 1/8 or 1/16 of a sheet. But the direction of the fibre should always be taken into account. The folding direction only makes a difference if you are using Nepalese papers or other qualities of papers.

Marking up

Using the set square and the pencil, mark the points you are going to cut along the upper and lower bows. The

Esther Fölmli (Switzerland): woven table runner made from cut and twisted shopping bags.

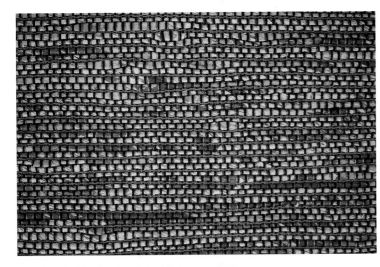

Monika Künti (Switzerland): braided room divider made from cut strips from old maps and architectural plans

Shifu experiments using different paper by Uta Heinemann (Switzerland)

Claude and Andrée Frossard (Switzerland): 'space and symbol',
tapestry 165 x 135cm, woven newspaper

distance you choose between these points will later determine how thick or thin the resulting thread is. Usually, the width of the strips varies between 2 and 10mm (1/12-2/5 in.). For your first attempt however, you should choose intervals of at least 5mm (1/5 in.) so that you do not run the risk of continually have to rip the paper strips off when you are twisting.

Cutting

In the next step, the sheet is put on a hard surface which you can cut on. Using a paper knife, cut it into strips along both of the points which are marked on opposite sides. Make sure that you cut through the central join on the inside, but that you leave an approximately 1cm (2/5in.)-wide, uncut, linking point where the outside edges of the sheet overlap. When the sheet is finally unfolded, it keeps its closed, right-angled shape, but on the inside there are many slits. When you use this method, even though there are a lot of cuts to be made, order is maintained and this is essential for the next step.

The purpose of folding is to shorten the necessary cut lengths so that you can work rationally and in exactly parallel. Experienced shifu-makers place several folded sheets over one another when cutting. However, if you do this, there is a great danger that the pile moves or that you do not completely cut through the bottom sheet.

Moistening

The unfolded, slitted paper is wrapped up in moist towels for seven to eight hours (usually overnight) and to prevent early evaporation, it is put into a plastic bag. This increases the elasticity of the threads for when they are made into yarn. Depending on the type of paper used, this step varies from being essential to less important. Japanese paper (*kozo*) cannot be worked on further if it has not been sufficiently moistened. Other paper, however, like the Nepalese *loktha* paper, can be worked on immediately after cutting when it is almost dry. For each paper, you have to

learn about its qualities from experience and discover the ideal amount of moisture which is needed.

Rolling

The next day, the slitted sheets of paper are rolled out on a hard surface such as a bamboo mat or porous stone. This makes the individual paper strips separate and crumple and begin to twist slightly. You will need to work carefully here, so that the paper is not damaged. Roll in the same direction with both hands, initially with slight, then

1) Folding and 2) Marking up

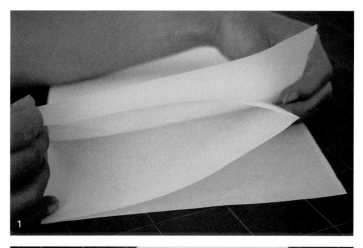

increasing pressure. From time to time, the paper should be shaken out to separate sections which have become caught up. You should then lay it out again and continue rolling. The link to the edges makes it possible to arrange the strips again and work on them regularly. Depending on the type of paper and evaporation, the material needs to be moistened using the spray from time to time. It is important that the paper should never be soaked through with water because areas which are completely wet easily rub: what you have learnt about quantities from your own experience is very important here too. After about 10 to 15 minutes, the paper is usually sufficiently pre-shaped and can now be divided.

Tearing
After rolling, the slitted paper is torn apart in such a way that an 'endless' band is produced. This is achieved by tearing the link on every second slit across the width of the paper through to the outside edge. This process is then repeated on the other side, but this time it should be

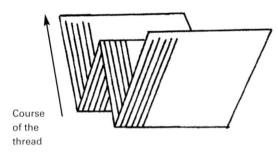

Course of the thread

staggered by the width of a strip to make a zig zag band. The crossing points should already have been twisted slightly between your thumb and index finger during the separation process. This is to stabilise them so that they cannot tear any further. Even after the thread has been woven, these turning points remain slightly thicker and this is how the shifu material obtains its characteristic bobbly structure in the finished weave.

The long, torn paper bands produce loose balls. These are put into a round plastic container which is then covered with a damp cloth. With some paper it is recommended that the material is covered overnight with a damp cloth again before it is twisted. However, if you are working in a room with quite a high level of humidity, you can move straight on to the next step.

Twisting
Pre-treated paper can be twisted into a compact thread by hand, with a hand-operated spindle or using a spinning wheel. This procedure is often incorrectly called 'spinning' or 'cotton spinning'. But here, individual loose fibres taken from a non-woven fleece are not worked into a yarn as is the case with spinning. Nor are several individual yarns joined to a strong thread (cotton spinning). The cut strips of paper are simply twisted into a thread.

Twisting by hand
If you only want to produce a small amount of shifu yarn and want to learn the basics of making threads, the only tools you need are your own hands! You should sit at a table and put the loose balls into a container under the table on the left. Then, with your left hand, bring the start of the thread up. Wet your right hand and roll the paper strips forwards, i.e. away from yourself. Once a small piece has been rolled out, hold the twisted end down with a weight (for example the spray bottle) on the right side of the table and start, working from right to left, twisting the strips using your thumbs and index fingers. Once you have

Folding and cutting diagram
▼ 3) Cutting

4) Folding
5) Moistening
6) Rolling
7) Pre-shaped paper strips
8) An endless zig zag strip is produced
9) The crossing points are twisted slightly during the tearing process.

finished a piece, weigh the yarn down with something on the left hand edge of the table and wind the twisted thread onto a square cardboard box on which a slit has been cut on each side. Now weigh the wrapped box down on the right hand side of the table, twist again, weigh down, wind and repeat the process from the beginning. When you are working, the loose ball on the floor should unroll too and it should not be allowed to get caught up. Shorter pieces can simply hang down from the left hand side of the table and can be twisted into a longer thread later by moistening them again. This method is ideal for smaller objects and for testing colours and paper quality.

Twisting with a spindle
A simple hand spindle consists of a (approximately) 30cm (12in.)-long, wooden circular bar with a hook on the top width and a pointed bit underneath. A few centimetres above the lower end there is a wooden cross or a disc (turntable) of 10 to 20cm (4-8 in.) in diameter. Like a gyroscope, the device can be set to a regular turning motion which is then passed on to the material to be woven or twisted. This simple apparatus has been used for

Tearing

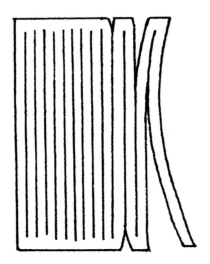

thousands of years in many different cultures for making threads.

To twist a paper thread using a spindle, a metre-long cotton, woolen, or similar thread is fastened to the bar underneath the turntable. Start to twist the end of the paper band by hand and then knot this to the stronger thread which you then take up diagonally across the bar and wind twice around the hook. Then you begin to turn the spindle loosely upwards so that its own weight tightens the thread. The twisting is transferred to the paper band and forms a compact thread. Material is gradually added until the spindle is almost on the ground. Just above floor level, stop the spindle, unwind the paper thread from the hook and the bar and wind the finished yarn underneath the turntable (or the wooden cross) in the same direction as the previous twisting. Bring the rest of the twisted thread up along the bar and around the hook and set the spindle into motion in the same direction and repeat the whole process. If you are using a very thin band of paper, it is advisable to set the spindle in motion on a table (as you would with a spinning top), so that the weight of the spindle does not weigh too heavily on the filigree paper. It is also important when you wind the finished yarn under the disc, that you never lose hold of the paper band. You should also maintain the tension so that if you overspin, this does not spread to the material which has not yet been turned. This is easy to do because of the hook. Because you constantly have to wind on, repeatedly interrupting the turning process, using a hand spindle, to make shifu yarns can be quite time consuming. However, with only a little technical help, you can produce very nice, regular threads.

Twisting on a spinning wheel
If you want to make larger quantities of shifu yarn, then the spinning wheel is best. When you press a pedal, a swinging wheel is set into motion and this motion is transferred to the material to be spun. Because the finished yarn is wound onto a wing spool during spinning, the

process does not have to be continually interrupted (as was the case with the spindle) and so it is quite quick.

You need a certain level of practice to be able to twist shifu yarn on a spinning wheel. You need to set the brake so that the yarn does not pull too quickly. Then you tie the beginning of the paper strip to a thread which is fixed to the spinning wheel, set the wheel into motion and start to push the pedals slowly. As with normal spinning, the material is added gradually and is pre-formed between the index finger and thumb. The strips are already creased from the previous rolling process and so once they are on the spool, their sides are irregular. All movements should be quite slow and cautious. The correct amount of moisture is very important for a good, compact yarn. If the paper is too dry, the fibres are inelastic and brittle. It feels 'wooden' and does not produce a nice thread. It is difficult to untwist a dampened shifu thread which has been worked on, back into a strip. But this is easy to do if the yarn is dry. Equally, if the paper is too damp, it is unsuitable for twisting as the paper is not very tear-proof when it is wet and untwisted. Areas which are too soaked completely dissolve in the hand and make the working process impossible. Washi absorbs water quickly and regularly and dries just as quickly. Therefore finding the right level of moisture is always a bit of a tightrope walk. By covering the material with wet cloths and occasionally spraying it on your hands, you can regulate it with a bit of experience and prepare the paper perfectly.

The finished yarns are incredibly strong and tear-proof after twisting and they have a beautiful surface structure which seems very alive because of the regular appearance of the burls. Every paper can be worked slightly differently and the character of that base material is then reflected in the twisted yarn.

Twisting using a spinning wheel

Industrially-produced paper yarn

Different types of paper yarn

The technology which was invented at the end of the 19th century in Germany to produce paper yarn has continued to improve over the last few decades. In theory, however, even the yarns which are produced today still follow the same principle as that of a hundred years earlier.

The *spinning paper* needed to make yarn is usually *sulphate* or *sulphite paper* whose fibres are not particular long. Because the cellulose used is often not completely open, it is very firm and stable but can also be produced in a very thin form. This is why it is often called heavy-duty paper. The strength of the spinning paper is usually between 15 and 60g/m. This, together with the width of the cut strips, later determines the thickness of the yarn which is produced.

Cellulose yarns

When the paper yarn industry began in Europe, the development of *cellulose* yarns was encouraged. In this process, the strips needed to make the yarn were not cut from the finished paper, instead it was attempted to make the strips directly on the paper machine. Certain devices on the sieve were supposed to separate the fibrous mass during the papermaking process itself. This then produced thin bands which could be twisted. From 1890, many patent applications were received and they all concerned this principle. The patents included Dr. Mitscherlich's method, the Kellerish paper yarn method, Kron's *Sivalin yarn*, and Türk und Issenmann's *cellulon yarn*. From 1890 to about 1910, cellulose yarn was seen as an important, forward-looking substitute material. It was gradually phased out as dry, cut paper yarns were introduced until its production came to a complete stand still in the 1950s.

Paper yarns

In 1908, Emil Claviez from Leipzig applied for a papermaking patent for cutting the finished paper into strips and then twisting them. These yarns were previously known as *xylolin yarns* or *licella yarns*. Almost all paper yarns sold today are produced in this way, or in a similar way.

With this method, the finished paper arrives at the spinning mills from the paper factories in large rolls. The paper cutting machines then cut it into 15-120mm (6/10-5in.) wide strips.

These bands are then wound onto disc, or the *turntables*, which is about 30cm (12in.) in diameter, and moistened before they are worked further. The twisting process takes place on a disc spinning machine which Claviez invented for this purpose. The paper discs are put into the machine horizontally and the twisting is started. The paper band is released from the outside edge of the disc onto a spindle where the spinning motion is transferred to the strips. Then they are twisted into a compact thread which is then wound straight onto the spool.

The advantage of this method is that the paper required can be produced from standard paper machines and no expensive equipment is needed. Although there are more steps involved in this procedure than with the cellulose yarn production method, the former method is more effective because only further processing needs to be carried out by specialised factories.

Blended yarns

It was also common in wartime to produced blended yarns with some paper parts. In the case of what was known as *textile-free yarn*, a fibrous gauze of glue and linen, cotton, jute or hemp rags was put onto finished paper. It was then cut and twisted. *Textilite yarn* was different. Either separate, loose textile fibres were added to the paper strips on twisting, or threads from cotton, hemp or jute which had already been twisted were wrapped around a paper strip. These threads were produced because of a lack of other materials and their priority was to imitate these. They are no longer important today.

Paper yarn in these special blends only continues to be used for a few things and then it is produced in large quantities. These are, for example, bindings or ropes for wine growing which are reinforced with a wire core, and products in the cable industry where the isolating effect of paper is useful.

Characteristics of paper yarns

Paper yarns are often not recognised for what they are and can be confused with sisal or raffia because they are sometimes used in the same way. However, if you look more closely, the material has very different, unmistakable qualities. It does not have a raw, fibrous surface like the majority of heavier-weight twisted yarns. It is slightly shiny and stiff. On the one hand, it has a natural and immediate effect, on the other, in its processed form, it also has a very stylish, functional character. Paper strings can be produced in various yarn strengths. Depending on thickness, the character of the end product is very different, but even the thinnest and most filigree-like yarns never lose their original body and tension with which inimitable effects can be produced.

In its twisted and woven form, paper yarn is extremely sturdy and tough. It can absorb up to 40% moisture, but loses a good deal of its strength in this way. Using various waterproofing methods, equipping with anti-mould products

Industrial paper yarns

and pressing, durability, shine and water-resistance can be considerably increased. With the right equipment, you can wash weaves in the washing machine at up to 30°C. Because of the thickness of the fibre, they keep heat in, but are still very light. The fairly smooth, somewhat slippery surface is dust-free and does not fluff. This is particularly advantageous for packaging material and is good news for allergy sufferers. Because paper yarn has a very long life, it is also an ideal material for use in the home.

Today, the ecological value of the yarns is very important to us. Paper comes from a natural, renewable source material and can be produced and refined in a relatively environmentally-friendly way. Once a product has served its purpose, it can be burnt or composted without releasing any harmful substances, or it can be recycled to make new paper. These are all qualities which are increasingly important today.

It is well-known that paper can be easily dyed and bleached and this also effects the yarns it produces. Sulfit paper is particularly suitable for this as it gives bright colours and is well-known for being colour and light-fast. If pigments are added to the cellulose mass before the papermaking process, then the paper produced is known as *spin-dyed paper*. Because the dye does not penetrate right to the core, different shades of colour are seen when the ends of the string are untwisted back into a paper strip. This exciting effect can be used deliberately in the creative process.

Trade-standard material qualities

There are only a few paper spinning mills in the whole of Europe today. Some traditional factories have resumed production on the old machines and modernised the process as a result of the current boom. The expensive machines make the heavy-duty paper rolls from the paper factories into yarns and deliver the raw material to various colour mills. There are big differences in the naturalness of the light, environmental acceptability and the depth of colour penetration, depending on where and with which pigments the material was treated. You can buy it all these days: from heavy metal yarn with its health risks, to high quality, even 'edible' yarn. You should check yarn quality very carefully, especially when working with children. The yarns which are found in Europe usually come from

Scandinavia, primarily from Finland. But there are also paper spinning mills in Germany, Austria, England and Hungary. Sometimes these products have a slightly different character to that of the Finnish strings. Various qualities of paper yarns are available in specialist shops and from most leading yarn companies. Their strength is given in the measuring unit Mn (metric number) and these tell you how many kilometres there are in a kilogramme of yarn. This value can be worked out by dividing the length of the yarn in metres by its weight in grams. The higher the number, the finer the quality of the yarn. The following yarn qualities can be bought:

- Simple twisted paper in various strengths:
  - The most widely distributed strength is Mn 0.8, a relatively thick, robust paper string. It can be obtained nearly everywhere and in many different colours. Depending on the company, it comes in 500g skeins or on spools.
  - You can also get a slightly thicker version of this yarn, which is just as brightly-coloured: (Mn 0.16) and the thinner version (Mn 1.65).
  - Very thin, simply twisted paper strings (Mn 7.5 and finer), which have a wonderful, almost graphic character, are currently only available in white and natural brown. Small quantities can easily be hand-dyed.

- Woven paper strings:
  - Thick, woven paper strings are only available from a few companies. These strings usually consist of four twisted yarns and they have a very robust character. These are available in strength Mn 0.32/4 in many different colours but the thinner variety (Mn 96/4) is only available in white and natural brown. A Finnish company also supplies colourful twisted strings, but this effect can also be achieved by hand.

- Paper bands:
  - Brightly-coloured paper bands are now available too. These are paper strips, folded slightly lengthways. They are irregular and made from the source product of the Mn 0.8 paper yarn before twisting. When it is worked, very natural, 'papery' textiles are produced.

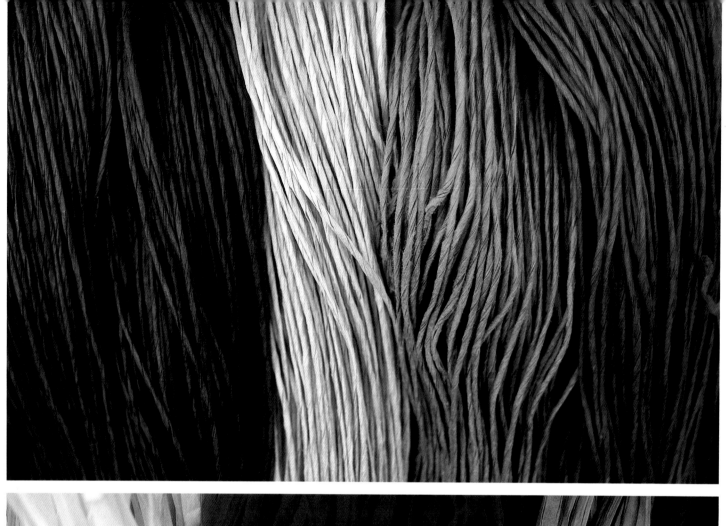

- Connective threads, wine cords, isolotaing material:
  - In gardening shops and DIY stores, you can often find cheap natural brown paper string. It usually has a wire core because it is mainly intended for use as binding cord for plant cultivation, or it is supplied as isolation material. Of course, it is not very high quality, but it is very well suited to experimentation.

Depending on the company and the amount ordered, all of these qualities are available as skeins, on spools, or cones. Craft shops also sell small packets. These contain the exact amount (already measured and cut to size) needed to make a small object, such as a paper flower, and also include instructions. However, these mini portions are expensive and limit creativity. It is best to get in touch with larger yarn manufacturers and keep your eyes open, because you can often come across material in many places where you wouldn't expect to find it.

Twisted paper strings

▲ Simple twisted paper strings
▼ Paper bands

3 Designing with paper yarns

▲ Design object
◄ Pile

The basics of textile design with paper yarn

Both types of paper yarns, Japanese shifu threads and European paper strings, are stubborn materials which are full of character. With their particular powers of expression, they inspire many different uses. Effects can be achieved which would be unthinkable with other materials. These continue to achieve surprising effects because the paper's uses as a textile are still largely unknown.

Before you start to make larger products, it is important to get to know the material using smaller samples and simple ideas which do not require great technical ability. You should also find out about its specific characteristics in order to develop a feeling for how the yarns react and what they are best suited to. If you use the threads as a textile material with its own, separate meaning, rather than a craft material, then you will continually discover new facets of this versatile material. You can then also use its very particular 'papery' language for your own creations.

Peculiarities when designing with shifu yarn

The shifu thread has its own unique character and because of its burls is structured in a natural way. When you are working with it, you should find a form whereby its specific meaning comes in useful. Too much time spent on the technical aspects and a over-ornate style are rather counter-productive because they draw the attention away from the material. Working simply is usually the best way to exploit its unusual qualities.

The shifu yarn is quite soft and therefore not very tear-proof. It is therefore best suited to techniques where it does not have to bear much tension, for example, knitting or as weft material for weaving. Shifu yarns are pleasant to wear next to the skin, they provide warmth and they absorb body sweat. The material is therefore well-suited to making clothes, but also jewellery and interior textiles. In particular, the bouclé structure of shifu textiles is clearly visible when backlit, in front of a window or a lamp. A pleasant, soft, yellowy light is produced.

Every shifu thread has a long history. The paper from which it was made gives evidence of the various plants, regions and people who contributed to it and it therefore acquires special qualities which are transferred to the thread. To understand the material, and be able to create a sound structure, it is important to know about these. Every sheet of paper has a slightly different colour, thickness, toughness, and mattness, and lets through a different amount of light. These differences can be taken advantage of. For example, in the same way as with rings on trees, the different shades of colour of a single sheet can be worked in a specific rhythm, to obtain subtle nuances and to bring the surface to life from the inside out. If you are making your own shifu threads, then through your choice of paper, strip width and the twisting power, the possibilities for influencing yarn quality are virtually unlimited. This is what makes this technique so interesting.

The burls which appear in the yarn at regular intervals show a definite rhythm which can leave their mark on the structure. If, for example, you decide on a weave width based on the size of the sheet of paper or you use the burls in your work according to a specific principle, you can produce thickened lines. Even the behaviour of shifu textiles when wet can be a starting point for design. When they are damp, paper threads are softer and tend to form waves and independent, expressive folds. If you deliberately use this independence and further emphasise its effect in the design process, instead of trying to control it, impressive shapes can be achieved.

Because of its long tradition, its entrenched religious and cultural importance in Japan's history and the enormous production cost, shifu thread is a very precious material. You should never lose sight of its background and it should be used respectfully and economically and only in places where it would make sense.

Pecularities when creating with European paper yarns have a much more unified and functional appearance than shifu threads. They have a smooth surface and can therefore depict all structures extremely precisely and clearly, producing an almost graphic effect (these appear blurred when rough, fibrous yarns are used). Because they are so regular and compact, they can be worked using more costly techniques and can bear a greater level of tension than shifu yarns.

Paper strings are very versatile and can be used in

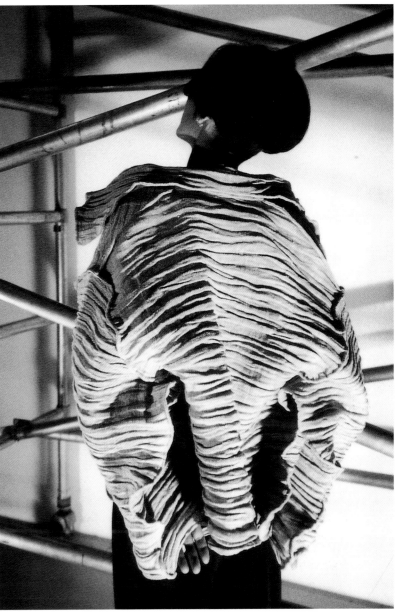

These square lights are made from iron frames. *Kozo* paper is stretched across them and decorated with shifu threads made from the same material. The burled structure of the yarn is clearly visible against a backlight and the paper produces a pleasantly-muted, yellowy light.

Esther Chabloz (Switzerland) made a series of clothes with a strong folded structure from Nepalese shifu material. Using a sewing technique, *Esther Chabloz* did not try to bring this natural folding under control. Instead, she emphasised it further by stitching outwards and allowing wave-like contours to form.

many different ways. Depending on the strength of the yarn, solid, robust textiles or almost floating objects with great elegance and ease can be produced.

The material is well-suited to winding, entwining and weaving as that is how it was first used at the beginning of its historical development. In the clothing sector, you can achieve many experimental results. Today, paper strings are used most frequently to make interior textiles and accessories.

Stiffness is a typical characteristic of this material and this makes it possible to achieve unmistakable effects. Even very fine strengths of yarn still have an amazing corpulence and inner stability. These can be brilliantly used to make three-dimensional objects. The high level of plasticity makes it possible to make filigree forms with loose gaps which could otherwise only be achieved with wire.

Paper yarns are, in theory, resistant and workable. They do however have some unique qualities which differentiate them from other materials when they are used.

- Because of their stiffness they are very difficult to work with and need quite a lot of strength to bring them under control. Thicker paper yarns are usually easier and more regular to work with when they are damp, although they are not so tear-proof in this state.
- Because the strings are not made of several loose individual fibres, but rather from one single, compact

▲ Even very fine paper strings (Mn 7.5) have a strange corpulence. Sonja Knoll (Switzerland) used this quality to make her crocheted object, entitled 'The heart of the Hieronimus Bush'.

◄ The smooth, stiff paper threads depict textile structures particularly clearly, as can be seen, for example with this board weave.

◄ A typical feature of the European paper string is its stiffness which makes it possible to achieve unmistakable, plastic effects. Paper yarns are well-suited to binding and entwining.
▼ To make this simple necklace, paper yarns (Mn 7.5 to Mn 1.65) were wound (with varying levels of thickness) around transparent plastic tubes.

▲ Weaves from paper yarns are stiff, but flexible and can be folded into various shapes in the same way as paper.
► These chains are made of short pieces of brightly-coloured paper strings (strength Mn 1.65) and strung together using a needle and thread. The stiffness of the string gives a coral-like effect.

paper strip, they are also extremely inelastic. They do not chafe and they do not lose their stability. Instead they either withstand it completely or they immediately tear apart completely. If smaller imperfections and irregularities in the tension are not balanced out, they multiply as time goes on. For this reason it is important to work very exactly from the very beginning. If a technique turns out to be suitable and everything is correctly adapted at the beginning, then there should not be any problems with the material.

- Because the paper yarn is twisted on one side and is not twisted in the opposite direction, it sometimes has a slight pull. If you twist a thread which is not taut, it usually forms a small loop. You should never pull both ends to try to get rid of this as you might do with other materials. If you tried this the area of weakness would tear straight away because of unsufficient elasticity. The best thing to do is to try to undo the little knot by turning in the opposite direction.

- Because of the smooth surface and stiffness, when the yarns are on spools or balls, they have no natural adhesion. Instead they spring up and have the urge to continuously unravel. To avoid the tiresome untangling process, the best thing to do is to put the ball or the spool into a nylon stocking which provides the necessary resistance. This makes all the difference when you are cutting a warp from paper yarn and it gives the yarn a regular level of tension and makes the yarn unwind steadily.

The colour and combination possibilities of European paper yarn

European paper strings are now available in many attractive colours, but all types of paper yarns can be dyed by hand just as well. Cellulose fibres absorb chemical dyes quickly and permanently. Yellowy or natural brown paper, in particular, produces rich, high-quality tones. To achieve these, the yarn is simply wound onto skeins, moistened and put into a soaking solution of textile dyes (trade-standard batik dyes) for at least half an hour and then put into a fixer. Water-soluble wood stains can also be used, although these are harder to fix. It is important to keep undoing the skeins as when the yarn is wet it pulls together

and wrinkles, making it even harder to untangle. Paper yarns, like all plant fibres, do not hold natural plant dyes well on a permanent basis. However, if correctly pre-treated, you can achieve particularly vibrant colours as shifu weavers from Asia have proved. The special recipes and passed-down traditions required for this are a science in themselves. Because the pigments do not completely penetrate the paper yarns in the dyeing process, the inside of the yarns remains lighter. If you untwist the ends of the paper strip yarns, you can see the various shades. You can deliberately use this effect in the creative process. The flag or flame-like borders which look almost painted can be used for decorative purposes. The untwisted yarn ends can, for example, be used as a stopper instead of a knot to prevent them slipping out. By using the combination of twisted and untwisted paper strips in one piece of work you can produce surprising effects and the paper aspect of the textile.

If you combine paper yarns with whole pieces of paper, this effect is even clearer. For example, it is possible to work little paper tiles into the weave, to wind paper yarns around sheets of paper, or stick them on to the paper. You can also introduce the threads into the sheet when you are making paper by hand to produce a relief-like surface. In this way, there is a direct reference from the thread to its source material.

If you combine the paper thread with other textile materials, you can also produce interesting effects. For example, you can use yarns with opposing characteristics in order to achieve an exciting contrast. Good examples of this are fleecy mohair, unspun silk and very elastic synthetic fibres. The unusual qualities of paper strings can be further emphasised when they are combined with materials which are stylistically related, for example, wire or horse hair which, like paper yarns, are stiff and smooth.

▲ All types of paper yarns can easily be dyed with chemical dyes. Natural dyes, as shown in this picture, produce wonderful tones, but they only last if the paper yarns have been correctly pre-treated. If you untwist the ends of the paper thread, flame-like borders appear and these emphasise the papery character of the materials.

▼ On these knitted wrist warmers made from shifu yarn, across a few stitches, a second, brightly-coloured yarn was introduced and their ends were twisted when finished.

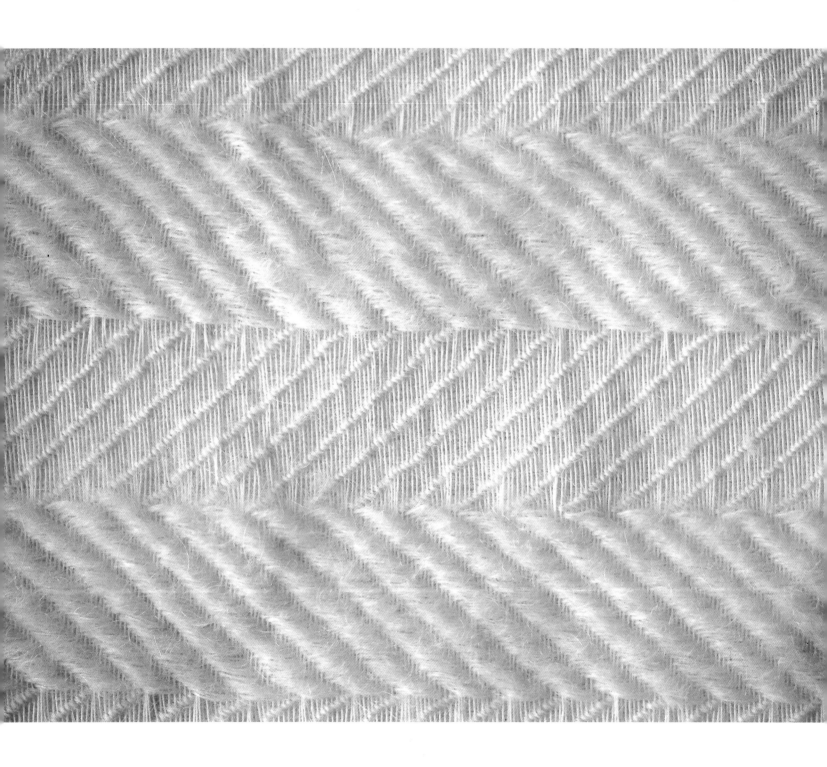

The combination of paper yarn (strength Mn 7.5) with fleecy
mohair produces an interesting contrast on this twill weave.

Paper strings and horsehair have similar levels of stiffness and they are both difficult to work with. For this weave, Veronika Rauchenstein (Switzerland) combined both of these materials and achieved an exciting interplay.

Textiles where the Japanese shifu thread relates to the European paper yarn are equally attractive.

The basics of creating products with paper yarns
Many traditional techniques can be used in different ways and reinterpreted with the various kinds of paper yarns. New ideas always emerge whilst you are working – the material inspires experimentation. You can try out completely different structures and let the material dictate the direction. Ideas for larger objects or related objects often develop from the beauty of a structure, its special haptic character, its colourfulness or the material quality of a surface. Whilst experiments often arise spontaneously and intuitively, you should always plan more precisely when you are making specific products which are to serve a definite purpose. Material qualities, measurements, colour and many other factors are connected and mutually influence each other. These should all be taken into consideration so that you obtain a coherent whole.

In this two-for-one weave, a woven layer is made by altnerating one shifu yarn weft with one European paper yarn in a warp of silver wire

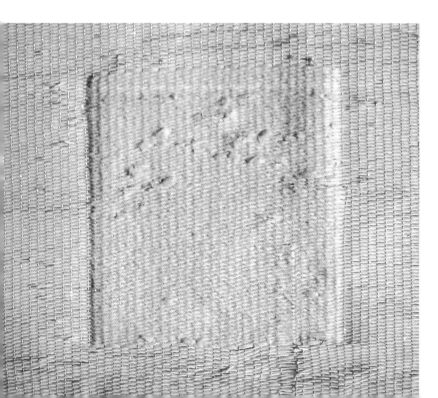

The following ideas and questions should get you thinking and may be helpful:

- Is the paper really suited to how you plan to use it? Is it functional for this (tear-proof, shower-proof, water-resistant, haptic etc.)? Can I achieve the desired effects, or the desired characteristics (better) with another material?
- Have I limited myself to the essential in the planning process? What are my basic design criteria (for example, unusual combinations of material, organisation surfaces using various structures, uniformity and variation, colour contrast etc.)? Can this main idea be developed or will it be eclipsed by other less important structuring methods?
- Which proportions am I choosing for my object? Can I refer to a module for the dimensions by saying, for example, that one length is an exact multiple of another? Does the thickness of the textile surface, the fineness of the structure or the size of the pattern fit together with the complete measurements of the object?
- Does the object fit into the environment for which it is planned? Do its measurements work with the space, does its colour match, can it, for example, relate to its environment by adopting individual elements of it?
- How do I structure the borders of my product? (Cleaning up selvedges, fastenings, seams, handles, bases, suspensions, specifications etc.)? Do I have to incorporate other elements? Will I find a solution as a logical consequence of the techique? Can I use standard ways of working with paper? (Cutting, using wood glue etc.) without losing the textile-like character?
- Is the end result worth the amount of time and money it will take to make it?

Gerlinde Fuchs (Austria) made white paper strings directly
from a cellulose pulp specially for this installation.

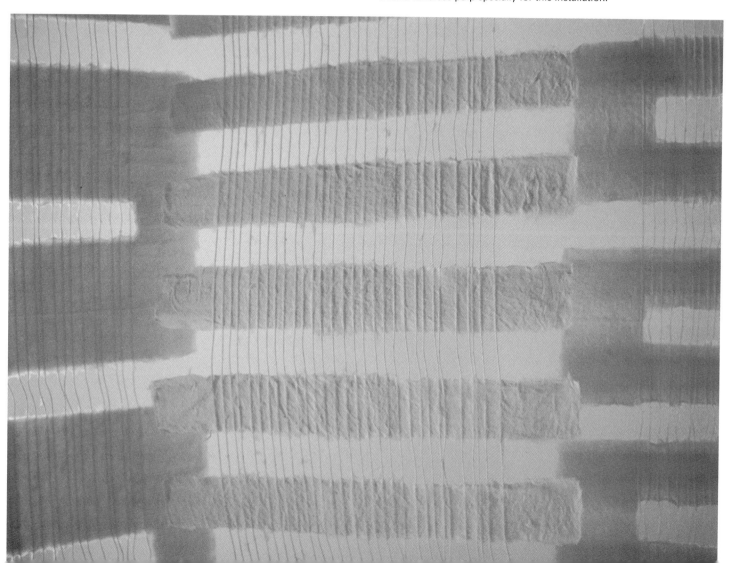

3.1 Entwining and knotting

In this chapter, I will be introducing techniques which produce a textile structure from just one continual thread system. In addition to the four primary entwining techniques which will not be described in more depth here, classic mesh methods like crochet, knitting and knotting a continual thread to make a net-like structure will be discussed.

Mesh is characterised by the fact that it is always composed of a thread which is continually tangling thread, the mesh, and can be separated by just one pull. Finished products made with mesh techniques are also called knits. In contrast to knitted or crocheted structures, net cannot be undone – they remain firmly knotted.

1 Christina Leitner
2 Marian de Graaff, the Netherlands
3 Christina Leitner
4 Marian de Graaff, the Netherlands
5 Maisa Turunnen-Wicklung, Finland
6 Cordula Hofmann-Molis, Germany
7 Katharina Frey, Switzerland
8 Uta Heinemann, Switzerland
9 Christina Leitner

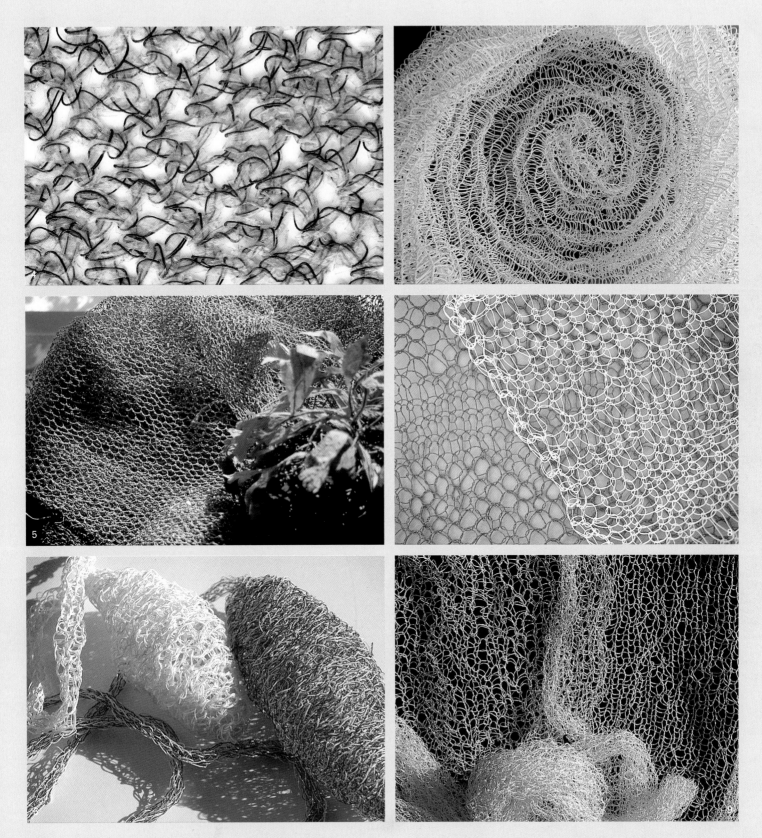

Knitting

Shifu yarn as well as the usual commercial paper yarn is suitable for knitting. The uniqueness of the shifu thread is particularly apparent in very simple, smooth knitwork. Stylish, soft structures with a matt surface can be produced and these feel particularly good when worn next to the skin.

Paper yarn and band also produce a very nice structure when they are knitted. They give a grainy, almost graphic surface. However, thick material is quite difficult to work with.

Ensemble

Materials needed:
- Shifu yarn made from *lokthah* paper (strip width of 5mm (1/5in.))
- Black Elasto-Twist (elastic yarn – cotton lycra mix)
- 5 knitting needles – strength 3
- Tight-fitting, black top

The ensemble consists of a cap and a top with knitted sleeves. Two different materials have been combined to form stripes on the sleeves. Six rows of Nepalese shifu yarn alternate with three rows of black Elasto-Twist. This material is much thinner and more stretchy than the paper yarn. This makes the knitted structure pull together at the filigree areas and a subtle, ripple effect is produced. The ensemble is pleasant to wear, keeps the heat in and can absorb sweat in summer.

The cap and the sleeves are knitted with circular needles and consist only of plain stitches. 132 stitches are cast on for the cap and distributed amongst the four needles with 33 stitches on each. It is made of a straight tube of 15 Elasto-Twist stitches alternating with 15 shifu yarn stitches. Then a further three rows of Elasto-Twist are added and tied off. The cap takes its shape from sewing and pulling together the upper part of the tube. On the back of the tube, the stripes are gathered together in a concertina effect, by folding together along the thinner, Elasto-Twist areas, so that a pretty cap shape which moulds the back of the head is produced.

The sleeves of the top are also knitted on four needles, although a total of 48 stitches are cast on, i.e. 12 stitches on

Knitted ensemble, design by Renate Egger, Austria

each needle. You should begin with the three, black Elasto-Twist rows and you should alternate about 31 blocks of Elasto-Twist rows with shifu yarn strips up to the sleevehole. The sleeve width is cast on in the last row of the shifu yarn strip, after the first stitch and before the last, in a rhythmic circle. No stitches are cast on in the first three shifu stripes. For the next nine, they are cast on at every third stripe and for the next ten, at every second stripe. For the final nine, two stitches are cast on at every stripe. Altogether there are 32 stitches on the needle.

Once you have stitched 31 blocks like this, you can begin to cast off so that you get a spherical arm shape. This consists of five additional blocks of alternating shifu yarn and Elasto-Twist strips. These can no longer be knitted, but consist of alternating smooth and inside out rows which both extend beyond the entire width. For the second row which is knitted with shifu yarn, different numbers of stitches are cast off after the first stitch and before the last:

once every three stitches, twice every two stitches, twice every four stitches, once every two stitches, five times every stitch, once every five stitches, twice every three stitches and once every two stitches. You should now have 12 stitches on the needle which can now be cast off.

The arm shape is approximately 63cm (25in.) long, with an upper arm width of 26cm (10in.) and a height of 12cm (5in.). Before you continue working it, you can place the knitted structure in warm water to make it softer. Then you can sew (by hand) the sleeves to a tight top, like the one in the picture, for example, which is made from black, elasticated velour.

The pattern for a top can be easily taken from a tight, well-fitting top. The measurements given correspond to a size 8 (UK).

Cushion covers

Materials needed:
- Paper band (strength Mn 0.8) in black, grey and natural white
- Knitting needles (strength 7)
- 3 cushion stuffings (50×50 cm)
- Silk lining in a suitable colour
- Sewing machine, sewing kit

These cushion covers are made of one smooth knitted square and one inside out knitted square from black, grey or natural white paper band. You will need to cast on 50 stitches to get a width of about 50cm (20in.). To get the same height, 50 rows are required. The finished knitted structures are very stiff, so you should put them in lukewarm water before you work further with them. Whilst they are still damp, you should iron them flat. This gives you a soft, fluid surface with an interesting, spacious structure and a slight shine. Both the squares should be sewn together (by hand) inside out on three sides using the thin paper yarn. Turn the cover over now, line it with a matching colour and sew the open side shut. The covers can be handwashed at any time, so these products are particularly well suited for use as compact floor cushions.

Cushion covers

Crochet

Curtain

Shifu yarn is not suitable for crochet because the dense stitched structure obscures the bouclé effect of the thread. Trade-standard paper yarn is, however, perfect for this technique because the material carries the structures of the different crochet variations particularly well. Because of the great stability and low elasticity of paper yarn, it is ideal for making compact products, such as bags.

Materials needed:
- Paper yarn (strength Mn 7.5) in white
- Crochet needle (strength 1.5)

This delicate curtain was made using a crochet technique from Turkey. This technique is very easy because it just consists of chain stitches strung to one another.

The example shown measures approximately 80 x 120cm (31 x 47in.) and was worked vertically. You should start with a row of chain stitches made up 450 stitches, then cast off, break the thread off and begin the second row beneath where you began the first. Here you should crochet one strong stitch in each of the first three stitches, then you start the loose chain stitches again. After the seventh stitch, insert the first row into the eighth loose section. Then crochet seven loose chain stitches and again link the second row with the corresponding section of the first. This rhythm should be kept up until the end, when the last three stitches should again be fixed to the first row. Then break off the thread and begin the third row underneath.

You should start with three loose stitches in the third row (they will later be connected to the ones in the fourth row). Then crochet just three chain stitches, insert the second row into the fourth chain stitch and then rhythmically crochet seven chain stitches. Each of these should be followed by a linking stitch to the previous row, as described before.

Because there were fewer chain stitches at the beginning, the fixing points opposite the previous row will each have been displaced by half a length. If you now alternate both versions of either seven or three chain stitches at the beginning, you will get a loose, diamond-like lattice structure.

The curtain in the picture consists of 56 rows. The bars were threaded through the first and last row and then hung on curtain hooks in the window frame. Because the crochet is under slight tension, you can pull the individual rows further apart or gather them together more. This produces attractive, transparent effects. The loose stitches look like raindrops running down the window. Have a look outside and let all the daylight into the room.

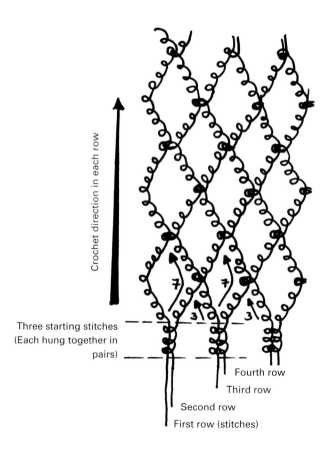

Crochet direction in each row

Three starting stitches
(Each hung together in pairs)

Fourth row

Third row

Second row

First row (stitches)

Stitch pattern for the curtain

Curtain

Pencil case

Materials needed:

- Tunisian crochet needle (strength 2.5)
- Paper yarn, strength Mn 1.65 in black, green, red and yellow
- Zip (20cm (9in.)-long)
- Sewing needle, sewing-silk

Tunisian crochet is a special form of netting and little-known or used in Europe. It is unique because it does not follow the pattern of one stitch after another being inserted, pulled through and cast off. Instead, in every row, all the stitches are cast on to the needle and only then, in the second step, cast off one after the other. Because there are so many stitches on the needle at the same time, this type of crochet is a little like knitting.

You will need the Tunisian crochet needle for this technique. It is longer than a standard crochet needle and has no handle so the stitches can spread out over its whole length.

You start Tunisian crochet as you would any other form of crochet – with a chain stitch. The stitches are cast on in the first row: as usual, you go into the first row and pull the thread through so that you should now have two loops on the needle. But instead of pulling the thread through these two stitches again, thereby reducing the number of stitches to one again, you go into the chain stitch and cast on the next stitch so that the number of loops is constantly increasing.When a stitch has been cast on in every chain stitch in the first row, you can begin casting off in the second row. The stitched structure should not, as it usually is, be turned over. You should keep working on the same side, i.e. the front.

In the casting off row, the thread is pulled through the first two loops on the needle, then again through the one just made and the following one etc. Starting on the left, the structure is loosened from the needle stitch by stitch until only the starting stitch is left on the needle. Now all the stitches in the third row are cast on and cast off in the fourth row etc until the required height has been reached.

Many mistakes are made at the beginning, when the number of stitches per row is not kept constant. The cast on stitches are automatically half a stitch width further

across than those in the previous row. This means, for example, that the number of stitches cast on in a row will alternate: first ten, then eleven, then ten again, then eleven again, then ten, eleven etc. These subtleties are somewhat awkward at the beginning, but if you keep a strict count, you will quickly get a feel for whether one is still missing in a row, or whether they have all already been cast on.

Tunisian crochet produces very thick, stitched structures with little elasticity. It has a simple surface structure, which is characterised by clear horizontal and vertical lines. Its appearance is therefore quite different to the character of other crochet techniques.

Used together with paper yarn, this technique is particularly suitable for making textiles which should be compact, shower-resistant and durable. You should bear in mind that you are limited to the length of the needle in the width of the structure you can make. It becomes increasingly difficult, the more stitches you have on the needle at any one time.

The pencil case shown in the picture has been worked according to the width. Only 18 chain stitches of black paper yarn have been cast on. You can see the coloured structure from the pattern. The yarn colour is changed after every row. The first row of chain stitches (and all casting off rows – rows 3,5 etc.) should be black. The rows in which the stitches are cast on (rows 2,4, 6 etc), the colours alternate in the following order: green, red, green, yellow, green, red, green, yellow etc.

To avoid the confusion of countless cut thread ends, the black yarn, which is always worked from left to right, should be pulled through the back of the colourful, cast-on rows to its new starting point so that it runs through the entire stitched structure. The brightly-coloured threads should be worked from right to left. At the beginning of a row, let about 50cm (20in.) of the new yarn hang down and then begin to crochet the row again normally. If the colour should be needed again, use the long end of the thread from the previous row. This means that thread ends are only on the left hand side and they can easily be worked into the base of the pencil case later on.

If you have never made a piece of Tunisian crochet, it is recommended that you attempt your first piece using just one colour of yarn so that you can concentrate on the number of stitches at the beginning.

The 11cm (4in.)-wide strip for the pencil case is 42cm long and consists of 112 cast-on rows. The beginning and the end are crocheted together at the finish. The ends of the threads are knotted on the underneath of the long side and crocheted off in black. On the upper side, a 20cm (9in.)-long black zip is sewn.

Bag
Materials needed:
- Crochet needle (strength 4)
- Paper yarn, strength Mn 1.65 in black and grey

The base and the sides of this bag consist of two separate parts. To make the side section, 105 chain stitches are crocheted and joined together to make a tube.Then you should crochet a round of tight stitches onto which rows of

▲ The stitches are cast on
▼ Pencil case

double trebles (they are wrapped around twice) are worked. At the end of this round, crochet three chain stitches so that you can begin the next row at the correct height.

The colour pattern is very simple. Three rows of black trebles alternate with three rows of grey trebles. Then another ten rows are crocheted in black and the side section is closed off with a row of tight stitches. The side of the bag consists therefore of 18 rows of double trebles each with a row of tight stitches at the beginning and the end, giving a height of about 38cm (15in.). To make the upper outside of the bag more stable, the border is reinforced with a row of crab stitches (tight stitches, inserted backwards, i.e. from left to right).

The base consists of raised tight stitches, i.e. a cover is formed, after inserting the stitch, pull it once through both loops on the needle. You start off with four chain stitches and lock them in a circle. In the next row, every stitch should be inserted twice, so that you get an area of eight stitches and the same in the next two rows. After the fourth row, you should have 32 stitches. From here onwards, alternate a row where you do not cast on, with one where you increase the number of stitches. In the fifth, seventh and ninth rows, cast on an extra stitch with every third stitch; with every fourth stitch in the eleventh row and with every sixth stitch in the thirteenth row. This gives you a circle of about 22cm (9 1/2in.) in diameter with an area of 105 stitches, the same number of stitches as for the side section. The two separate parts can now be easily crocheted together with strong stitches.

The handle is also made of strong, raised stitches. Ten rows are cast on and a total of 105 rows crocheted, giving a length of about 95cm (37in.). The ends are crocheted on to the upper edge of the opposite side section with strong stitches and the side edges of the handle are reinforced with crab stitches.

Bag

Knotting

Making knotted nets has a very long tradition. Originally, the technique was carried out without any tools, but over time, the net needle came into use. It carries the thread when the mesh is being formed: it loops itself around what is known as the roller, which ensures that the stitches are uniform in size. Although mesh offers some very exciting creative possibilities, today the process has largely been forgotten in the West.

Shifu yarns are not at all suited to meshwork because when the knots are tightened, a lot of pressure is exerted on the thread. Industrial paper yarns, on the other hand, give very beautiful, distinct knot structures. Paper yarn mesh is so attractive because of its stability which makes it possible, in spite of the very open nature of the structure, to make stiff, corpulent textiles with incredible ease. Because the paper yarn is stiff and its surface slip-proof, the mesh knots can only be pulled together in a regular shape with a certain amount of practice. If you have never had any experience with this technique, it is recommended that you do a few test pieces with another material (ideally heavily-woven cotton yarn), before you move on to work with paper yarns.

Net bag
Materials needed:
- Mesh needle (about 17cm (7in.)-long metal needle with slits at both ends); available in needlework and wool shops
- Mesh roller (about 15cm (6in.)-long, rounded off wooden or plastic strip, in 1cm (2/5 in.) length (available in needlework and wool shops)
- 2 wooden, bamboo, or metal hoops (with a diameter of about 15cm (6in.))
- Paper yarn, strength Mn 1.65 in various colours (hand-dyed shades of yellow, red and brown)
- Screw clamp

Before you can start knotting, you need to wind the paper yarn in the colour you want to start with, onto the mesh needle. You should pull it tightly alternately through the upper and lower slit about twenty times. About 50cm (20in.) of loose yarn should be left hanging off the needle and then the yarn is cut off.

Now hang one of the hoops from a screw clamp and

net

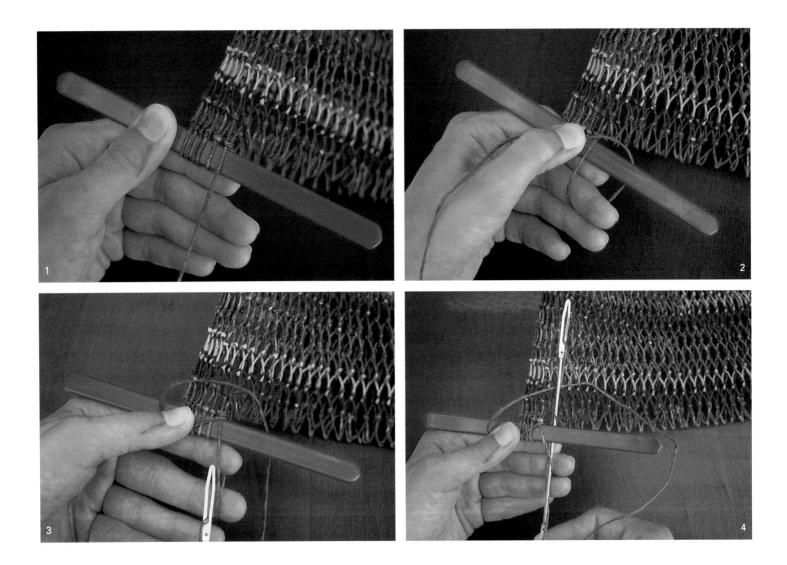

1 The thread is laid over the roller and the left hand.
2 The left thumb holds the thread tightly
3 The tread is placed around the roller in a half loop
4 Put the mesh needle through the loop produced

fasten the end of the wound yarn to the hoop so that you can make the first stitch. Take the roller in your left hand, hold it underneath the knotted thread on the outside of the hoop and, with the needle in your right hand, make a mesh knot over the roller. Pull the thread over the roller and your left hand in towards your body (1). Then take it down underneath your hand in the direction of the screw clamp so that the needle appears above the roller. Pull it up to the surface and hold the thread tightly on the mesh needle with your left thumb (2). Now take the needle along the underside of the roller again, towards your body, so that the thread you are holding makes a half bow around the roller (3). Then, put the loop, which has come from underneath, onto your left hand and feed the needle between the hoops and the bow up to the surface (4). Now pull the knot tightly by pulling the needle towards you and slowly removing your left finger from the loop. This makes the thread tighten around the roller and a knot is formed on the outer side of the hoop. Now you can start the next stitch. If the thread becomes too short, you can simply add yarn from the needle. As you continue, more and more parallel loops are formed on the roller. Then you just pull the roller out, leaving just the last few stitches inside it, and continue with the netting.

To make the bag shown in the picture, 60 starting knots were put on the hoops. Once the first row is done, then you can pull the roller out of the stitches completely and turn the hoop onto the other side, so that the needle is then hanging from your right hand. The roller is then held underneath the loop which has just been formed and you can continue with the next row. The knots are now gradually hung from the loops of the previous row, instead of over the hoops. At the end of the row, the hoop is turned again and the process continued, so that a net-like structure is produced.

If the yarn on the needle comes to an end, then you just wind the new colour on and join the ends with a figure-of-eight knot very near the last knotting point. Although the colour sequence is always the same in the bag shown in the picture (dark red, orange, brown, violet, apricot, yellow, dark red etc), the length of the wound yarn varies. This means that the joining points move every time resulting in a choppy, organic coloured structure.

If twelve rows are meshed on the first hoop, the same process is repeated on the second hoop. Now just the two single starting points are meshed together. After the twelfth row, the hoop is turned and the second hoop is also hung from the screw clamp. The thirteenth row is then put on the net which is now uppermost, both hoops are turned and the next row on the other section of net is continued. From this point on, you should be netting all around so that both sections are joined at the sides.

The bag takes its shape simply from the elasticity of the net stitches, no additional stitches are cast on. When they get thicker on top of the hoop, they stretch downwards more, producing a trapezoid-like shape. In this way, very individual, organic bag shapes can be created.

The bag in the picture has 36 rows. At the end, both of the loops, which were lying across one another, are closed at the front and back and a final row is knotted.

The future

'Endless'

The Swiss artist, Uta Heinemann made these three sophisticated, round objects from natural-coloured paper yarn. The insides are fitted with a wire core and are therefore extremely stable. The objects were made with no technical tools, using the primary entwining technique of simple hanging. Going out from a circle, a very simple textile structure came into being round after round. It built up organically and conquered the space like waves. The three plastic objects can always be rearranged and pulled out of shape. They play with the interior and exterior, positive and negative, light and shadow and playfully look at the possibilities of the technique and the material.

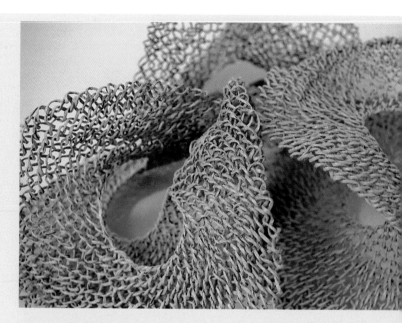

'Endless', Ute Heinemann

Shifu collar

Textile designer, Lis Surbeck, used this exquisite shifu collar to test out how very thick shifu yarns react when you work with them. She twisted the threads for the shawl herself using a hand-held spindle. Three large sheets of *kozo* paper were needed and she cut each piece into 1cm (2/5in.)-wide strips. This gave her a strong yarn which she knitted on thick needles, using only right stitches. The structure is 35cm (14in.)-wide, but has only 45 cast-on stitches. The shawl is coarse-grained and robust with a sturdy surface structure and the inside still keeps you warm.

Shifu collar, Lis Surbeck

Spiral container

This delicate container was made by the Swiss designer, Katharina Frey. It consists of a long tube knitted from thin, white paper yarn. It was made using a simple circular knitting device, the 'knitting board' which has six teeth. Nearly three metres (10 feet) of this tube were needed to make a basket-like container with a height of about 12cm (5in.) and 14cm (5 1/2in.) in diameter. Katharina Frey wound the long structure tightly together in a spiral arrangement and fastened the end with a feather. The different layers stick together automatically because the filigree spun yarn becomes entangled. The marvellous structure of this container is particularly clear when you look down into the inside from above. Because the paper thread is not twisted and is always knitted in the same direction, the tube has a slight pull. This tendency is further emphasised by the spiral arrangement producing an organic structure.

Spiral container, Katharina Frey

Jewellery

These various pieces of jewellery were all made from fine paper yarn (strength Mn 7.5) and some of them were hand-dyed with bright colours. There are different types of tube-like necklaces which were made using a simple circular knitting device with a large diameter. There are also various crocheted accessories. Simple, loose chain stitches in different colours produce pretty effects. Even the necklace, which consists of small, brightly-coloured circles is made using the crochet technique. The individual segments made from sturdy stitches were arranged in the sequence of the colour circle and fastened with pearls on sprung steel wire hoops.

Jewellery, Christina Leitner

Light object

This wind light was made by Mäti Müller from Switzerland. It consists of a 50cm (19 1/2in.)-long, smoothly knitted tube made from shifu yarn which is pulled and gathered over a 30cm (12in.)-high glass cylinder. By doing this, the material produces beautiful, soft folds. Mäti Müller spun the shifu threads herself from *kozo* paper. If you light a candle inside the structure, a pretty effect is produced, because the bouclé structure of the loose knit is particularly visible against a strong back light. The colour tone of the material spreads a pleasant light and the candle also casts an interesting star-shaped shadow on its surroundings.

'Circle of life'

This miniature textile object is made of nine pieces of square mesh made from thin, white paper strings which have been hung up vertically behind one another. There are differently-sized circles made from thin, white *kozo* paper on each piece of mesh. The net-like structure can still be seen through them. The circles are arranged according to size – the smallest on the outside, the biggest in the middle. The interaction of the individual layers gives the impression of a ball and the object almost appears to float.

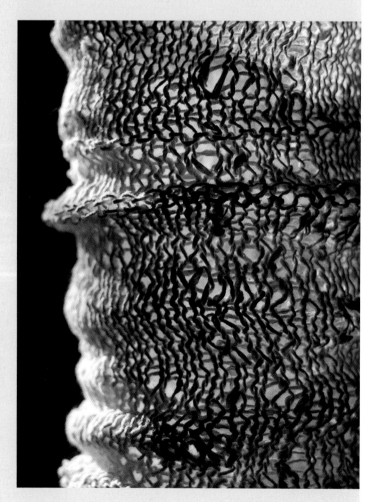

Light, Mäti Müller

''Circle of life', Christina Leitner

Scarf

The peculiarity of this simple piece of knitwork is the thickness of the needles in relation to the strength of the yarn. A very loose, holey, stitched structure is produced. In no way is it feeble. It demonstrates great inner tension. Because the thread is very stable, the individual stitches do not slide together. They take their place and produce an almost graphic line. This gem is extremely light and looks particularly decorative against a dark background. It can easily be draped to form a sort of turban because the stitches become entangled, giving it body.

'Giant's socks

These 'Giant's socks' are by Sigrid Schraube. They are made from white, knitted paper string, and are about 3m (10 feet)-long and 75cm (30in.)-wide. They were made for an exhibition honouring the Brothers Grimm. The German textile designer has been working intensively with paper and using it as a textile for several years now. With this huge installation she wanted to put herself with the giants and object to the enormous amount of time and material devoted to our fast-moving lives and the significance of material consumption in our time.

Scarf, Christina Leitner

'Giant's socks', Sigrid Schraube

3.2 Braiding and twisting

Braiding is one of the oldest textile techniques and has, over the course of history, produced a great variety of variations. Some of the most important sub-groups are winding, twisting, beaded partial braiding, circular, parallel braiding and diagonal braiding as well as all tube, cord and circular braids.

It is characteristic of all braiding techniques that there at least two groups of thread which can meet in different ways. In contrast to weaving, none of the thread systems are rigidly fixed to a frame, all of the parts are flexible. It is known as partial weaving, if only one of the thread systems is actively worked into the second, passive system.

The sprung technique, which is also called the Egyptian braiding technique, does not really come under braiding, it is a warp material method. A fixed, taut warp is twisted into itself to make a flexible, textile structure. A second thread system is not necessary for this type of textile, but it can be inserted to make the structure more stable.

1 Christina Leitner
2 Christina Leitner
3 Christina Leitner
4 Anita Dajac, Switzerland
5 Christina Leitner
6 Christina Leitner
7 Luzia Fleisch, Italy
8 Christina Leitner
9 Christina Leitner

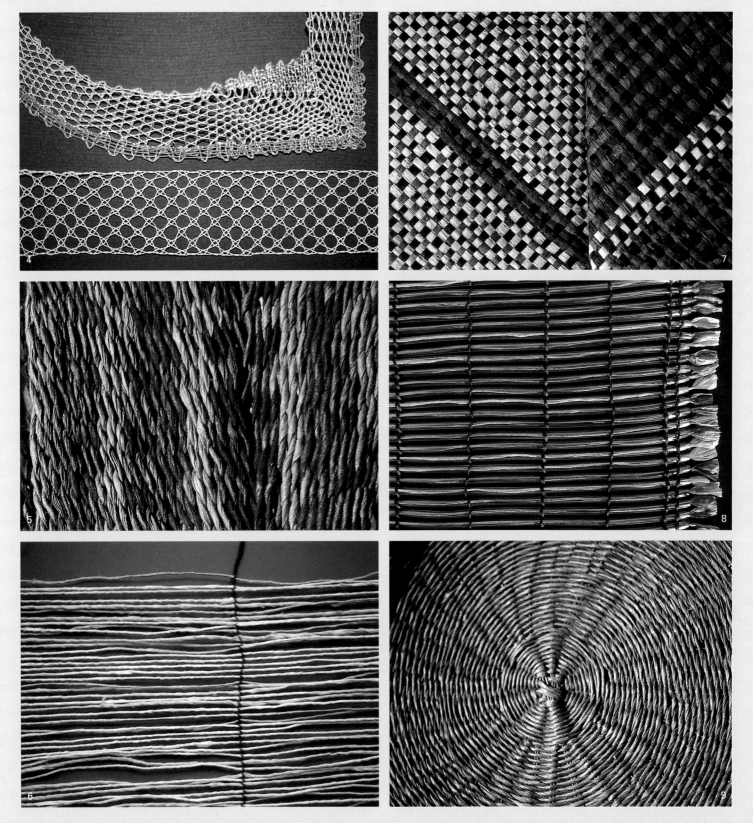

Partial plaiting

Partial plaiting means all techniques where an active thread system is worked into a passive one by binding or winding around it. Over the course of history, many different variations were used and are still used today, particularly in basket weaving.

Because of its stiffness and body, European paper strings are not dissimilar to traditional basket weaving materials such as rush, willow or reed. They are therefore perfect for these techniques. If they are woven into a basic framework made from wire or wooden bars, a nice, clear structure is produced. An increasing number of basket and stool weaving factories have recently become interested in paper strings because they represent a cheap, easy to work, alternative to natural materials which are increasingly difficult to get hold of. The bright colours and the fact that the length of the raw material is not limited by natural plant growth open up new possibilities.

Shifu yarn is unsuitable for partial braiding because the material has to be pulled tightly in this technique, should be as regular as possible and because large amounts are usually required.

Baskets

Materials needed:

- Circular rolls of steel wire (diameter of approximately 0.8 mm)
- Small combination pliers
- Paper yarn, strength 0.8 in dark blue and green
- Marker pen
- Ruler

These stable, round baskets with a diameter of about 22cm (8 1/2in.) are made from paper string woven into a wire framework. Different sorts of wires can be used for this and they can be bent into the shape desired. However, the diameter should measure at least 0.8mm, so that the structure is stable enough. To make the baskets in the pictures, a steel wire was used which was wound into a curve. This produces the half-ball-like shape of the basket.

Round sections about 50cm (20 in.)-long are nipped off the wire. This corresponds to slightly more than a half circle in the given winding process. A total of 32 pieces of this curve are fastened together in the middle and on the vertex with one half of a piece of wire wound round several times. The surplus end is nipped off. A total of 53 half curves start from this fixed point. They are then fanned out to form a three-dimensional basic framework into which the paper string should be woven, starting in the middle.

The difficulty with weaving from the centre out is that the radius in the base area is much smaller than at the outer edges. Because the proportions between the two systems should be very similar on the whole basket, with a regular thickness, the wires are only grouped into a few branches at the beginning and then gradually fanned out. Begin at the very bottom by separating the wires into nine groups of seven wires in the form of a star. Once you have bent the nine branches open into a circle. you can begin weaving the paper yarn alternately above and below the wires. It is important that you pull tightly whilst doing this and that you always push the string towards the centre to produce a regular structure.

If you have come back to the starting point after one round, the process is simply continued in a spiral shape, so that row by row the inside of the base is made. With this type of curved weaving, it is important that you always have an uneven number of branches, so that after every round the position of the paper yarn above or below the wire is automatically exchanged, giving a plain woven structure.

Once you have done 32 rounds like this, the intervals between the branches will become increasingly large and the wires should now be fanned out for the first time. The nine branches are distributed into a total of 19 groups with three or four wires in each. They should be spread out so that one group of four alternates with two groups of three, leaving three groups of three over at the end. Weaving is then continued above these branches. In the first round, you need to hold the yarn particularly tightly so that the wires fit into their new positions as well as possible, After 38 rows, and once an average of about 12cm (5in.) has

▲ At the beginning, the wires need to be divided into small numbers of branches.

◄ To be able to work a nice border, the ends of the wires are turned up.

▼ Baskets. The instructions are for high baskets

been achieved, this distribution will also become too thin and the wires need to be divided up again.

They are now divided into nothing but groups of two and groups of three and the basket has a total of 31 branches. If you weave a further 42 rounds into the newly-made framework, you will reach the maximum radius of the basket and the wires are divided up for the last time into a total of 63 individual arms. Now the natural curved shape of the wires along the side can be woven upwards. Once you are about 5cm (2in.) from the planned border, you can start thinking about how to make it. To achieve this, mark (using the ruler and marker pen) the planned border point on each wire, keeping the same distance to the edge all the time. You should weave all the wires about 4cm (1 3/4in.) above this point and bend them up at the required spot. By turning the basket upside down, you can check whether the height is more or less the same all over and whether it gives a regular, flat border. If not, you need to correct something. At the very end, the bent wire loops are useful for a nice border edge. You should continue weaving now and enclose the two-track wire once you have reached the required height.

The last 3.5cm (1 1/2in.) are worked in two colours. The long stripes on the outer edge are produced by changing the colours on the surface. It is the same effect as with a weft rep structure. Whilst the yarn has been running alternately in front of and behind the wires, the second yarn is now brought into play. It is worked simultaneously, parallel to the first. The wires do not alternate each round, they are woven together, by alternately running in front of or behind the wire. They then cross and swap positions for the next wire. In basket weaving, this technique is known as lacing, and the weft is virtually the same as in twisting. So that the pattern does not move after every row, it is important that the number of wires should now be even. Therefore bend any two wires together and braid them together from now on so that the number of branches in the last section is reduced from 63 to 62. Because you work in twice as much material as with normal weaving, and because the wires are doubled on the upper edge, a very stable, compact border stripe is produced. Once you have gone up to the uppermost edges of the bent wires, you can make an edge, by pulling a paper string end through all the bent wire loops and cutting off the surplus ends. Thanks to this simple trick, the woven structure can no longer separate from the base material and a discreet, pretty border is produced.

Many different variations on the basket can be produced using this weaving method. You can weave flat bowl shapes from straight wires, make oval shapes from an elongated starting point or make big baskets with thick, paper yarns.Because of the flexibility of the wire, there are no limits to the shapes you can make. The appropriate proportions between the intervals of the base framework and the thickness of the paper string are the deciding factors for a satisfactory result.

Centre: 63 wires	
9 branches each with 7 wires after18 rounds	‖‖‖‖ ‖‖‖‖ ‖‖‖‖ ‖‖‖‖ 7 7 7 7 ...
19 branches from 3 or 4 wires after 38 rounds	‖‖ ‖ ‖ ‖‖ ‖ ‖ ‖‖ ‖ ‖ 4 3 3 4 3 3 ... 4 3 3 3
31 branches each with 2 wires and one lot of 3 wires; after 42 rounds	‖ ‖ ‖ ‖ ‖ ‖ ‖‖ 2 2 2 2 2 2 ... 3
63 branches from one wire; after 32 rounds	‖ ‖ ‖ ‖ 1 1 1 1 ...
Felt with two colours over 62 wires; after 22 rounds.	‖ ‖ ‖ ‖ ‖ 1 1 1 1 ... 2
Border	

Boxes

Material needed per box:
- a wooden board: 40 x40 cm (15 1/2 x 15 1/2in.), strength 1.8 cm (3/4in.)
- 32 wooden circular bars, 27cm (10 1/2in.)-long, diameter 1cm (2/5in.)
- 4 wooden strips with a square cross-section (1.8 cm (3/4in.) length of side), 2 pieces 40 cm (15 1/2in.) long, 2 pieces 36.4 cm (14in.) long
- Wood drill, saw, glue
- Paper thread, strength Mn 0.32/4 in 2 colours, for example red and violet, or blue and green, about 50m (54 yards) per colour.

The base of these compact boxes consists of a square wooden board, 40cm (15 1/2in.) in length along the side, into which about 32 holes each with a diameter of 1cm (2/5in.) have been drilled at regular intervals. The interval between each hole is barely 5cm (2in.) and 0.5cm (1/5in.) to the outer edge. There are therefore nine holes on each lateral edge. Holes of the same depth are also drilled into the wooden strips which are to be used as a border.

Wooden circular bars are now inserted and glued into the (approximately) 1.2cm (1/2in.) deep holes, to produce the base grid for weaving. The bars are 27cm (10 1/2in.)-long, so the proportion between height and length of the side of the box corresponds to the golden rule (a ratio of 1:1618). This harmonious proportion should mean that the large boxes do not appear bulky or heavy. The uniform character is further emphasised if the basic framework is painted with woodstain in a matching colour before weaving. There is then no contrast between the wood and the textile material and it is discreetly incorporated into the weave.

The woven sides of the box consist of two colours with similar character. The yarns are woven into the grid bars in rep weave. The colour is changed after each round, so that if there is an even number of grid bars on the outside of each row, the same colours should lie directly above one another, producing the typical lengthways striping.

Begin by knotting the two yarn ends together and tying the beginning around one of the corner bars, so that the knot is facing the inside of the box. The first yarn (violet) begins the weaving. It runs from left to right alternately in

▲ Using two yarns, a plain **woven structure** is produced.
▼ Boxes

front of and behind the grid bars on the first side. Once the first side is finished, the second yarn (red) is started a bar later using the plain weave. Once both of these yarns have run through the first side, the box is turned 90 degrees and both the yarns are woven into the second wall with the corresponding rhythm. By turning again, you get the next side wall to work on and the process is repeated. At the first side wall, automatically weave the colours correctly in the initial row. After several rounds, the colourful lengthways stripes appear.

The thick, bulky material is easier to work with when it is wet. If you simply pull the yarn through a water bath just before weaving, and then unwind it, you can get a really regular, close structure. You should be careful not to pull the damp yarn too tightly and do not let it become so taut that the bars start to bend inwards. It is best to check after a few rounds whether the border strips can still be easily put onto the bars.

Once you have woven 57 rows and reached a height of 19cm (7 1/2in.), you can release the upper section from the two opposite handles. You then need to work the upper section in two separate parts. Weave one side up to the fourth bar, then tie both yarns around this bar and work in the opposite direction up the fourth bar on the opposite side, at which point you change direction again. So that the yarn colours stay in the right order, you need to cross them over once after you have tied them onto the bar where you change direction. After 16 rows, a further 6cm (2 1/2in.) is woven and the total height is reached. Both of the yarns are cut off, the ends are knotted together inside the box and the second side section is formed in the same way. This produces two rectangular handle shapes which will be given a structure by the freestanding central bar.

Once you have finished weaving, put the border strips onto the ends of the bars. Both of the longer crossbeams should seal off the side walls with the handles. The strips need to be well glued because a good deal of pressure will be exerted on them through their own weight and the content of the box when it is lifted. Because the yarn is so thick, you do not need very much time to make these very robust boxes. The material cost for yarn and wood is, however, relatively high.

Twisting

Twisting is a special form of partial weaving. Two or more active thread systems are twisted together and set the threads of a further, passive system, by wedging this last system inside the twists. The linking points are very stable and are difficult to move. It is therefore possible, using the

Material for two boxes

twisting technique, to weave the 'weft material' together with only relatively few rows and to still obtain a compact surface.

Paper yarn is ideal as padding material for this technique. If you want to achieve a mat-like character, you can also use it as a robust binding thread for the twisting process. Because there is no pressure exerted on the 'weft yarn', twisting is one of the few weaving techniques for which shifu yarns can also be used. If you twist them in parallel, you can produce nice structures which show the burls on the material off to their best advantage.

Table sets
Materials needed:
- Per set, about 18m (20 yards) of transparent synthetic skin with an outside diameter of 0.5cm (1/4in.) (available from DIY superstores
- Sticking tape
- Water-soluble OHP marker
- Transparent nylon thread (diameter of about 0.35mm)

- Paper yarn, strength Mn 0.8 in both colour families: yellow, green, turquoise, blue and yellow, orange, red, pink.

To make these table sets measuring from 28 x 40cm (11 x 15 1/2in.), the paper string is pulled into artifical skins and is therefore just there for effect. For each one, you need 60 pieces of skin with a length of about 30cm (12in.). These are then hung from one another with a transparent nylon thread using the twisting technique. So that you can work precisely, it is best to stick the somewhat stubborn skins with sticking tape in parallel on a background and mark faint lines on them using a washable marker pen and a ruler. These lines show you where the binding thread should run later To make the sets in the picture, there are six parallel-running lines with a distance of 5cm (2in.) between each.

Start twisting at one of the middle lines and work step by step towards the outer edges. To do this, remove the sticking tape on one side. Cut off a piece of nylon yarn

Table sets

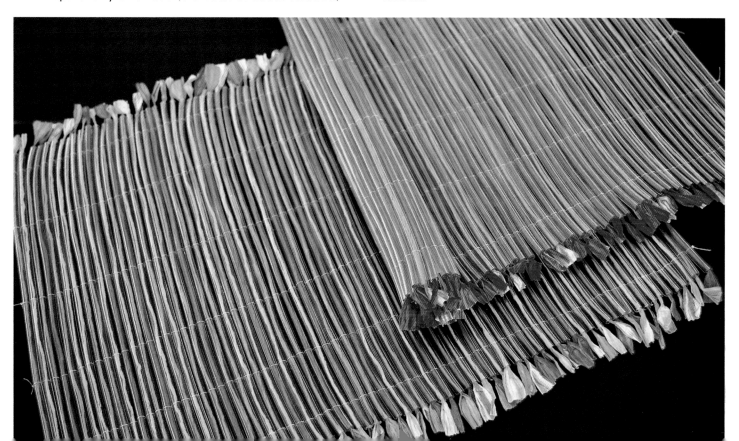

(approximately 1.5m (1 1/2 yards)-long), halve it and fix the centre of it with a knot. One end of the thread is then pulled through the first skin, just underneath the marked point. It is then crossed with the second, thread running on top so that the first thread can travel down in the second skin and they can swap positions. For the third skin, the first thread goes down again, for the fourth, the second thread goes down, for the fifth, it is the first thread again, and so on. Once you have hung skins together across the entire width, knot both of the thread ends and cut the surplus material off. You can now begin the next row. To avoid any pull, it is advisable to continue crossing in the opposite direction. For example, if you twisted the nylon threads together in direction s in the first row, the parallel lines should be in direction z. These twisting directions change over after every line.

Working from the centre out, once you have worked all the connecting lines to the outer edge, you can remove the second piece of sticking tape, turn the set over and complete the second half. If you make sure the twisting is just parallel to all the marked lines, you can wash these marker points off and cut the skin ends straight. If the surface does not lie completely flat, cover it with a slightly damp cloth and carefully iron over it. The steam smooths the synthetic material, but you must be careful not to let the temperature get too hot so that the nylon threads melt.

The sets are given an added decorative effect by inserting the paper strings. The pieces of thread in both of the colour families green-blue-yellow and red-pink-yellow are about 4cm (1 1/2in.) longer than the skins. Pull them through the hollow synthetic material in an irregular striped sequence. One end of each of the pieces of thread is twisted up and the other is pushed through the skin. Because the material is so stiff, this step can easily be managed without any additional tools. At the end, the threads are cut to equal lengths and the upper ends are twisted up to stop them slipping out.

This method produces pretty table sets which can easily be wiped down. They have a completely different effect depending on the background you put them against. They are very compact and dirt-resistant, but they also have a delicate side, because, at first glance, you can only see the loose, brightly-coloured strings which lie well-protected and embedded in their covers. If you like, you can also change the filling material without too much effort.

Diagonal braiding

In diagonal braiding, both thread systems are alternately active. They are characterised by the fact that the direction of the braid does not run parallel to the edges of the textile structure, but usually at an angle of 45 degrees. The crossing over possibilities are similar to the weaves in spinning. Although the basic principle of diagonal braiding is one of the simplest technical methods, its variations demonstrate highly complex techniques, as cultural and historical examples of sophisticated hair and cord braiding from different continents undoubtedly prove.

Shifu yarns are not suitable for diagonal braiding, but all types of paper yarn are, paper bands in particular, with which you can produce thick, multi-purpose surfaces.

Paper yarn in skins gives a colourful effect.

Cases

Materials needed:
- Paper band in strength Mn 0.8 in white, grey, black and red
- Black endless zip
- Transparent sewing silk
- Sewing machine, sewing kit

Both of these cases are made from just one braided strip which has been transformed into a three-dimensional object by skilfully sewing a long zip onto it.

The braided strip for the tubular brush case is 5cm (2in.)-wide and 125cm (48in.)-long. You need one white, four grey and four black paper bands, each about 4m (4 3/10 yards)-long. These are pulled over a fixed mounting in the colour sequence shown in the picture and halved in the middle. Cross the adjacent paper bands over in the middle. They are braided together in plain weave at a 45 degree angle to form what will be the outer edge. They go down the entire width until they reach the outside edge. They turn there, to run inside diagonally again in the next row and finally across to the opposite edge. In the same way as a three-way braid, the active paper bands alternate, whilst the rest functions as a 'chain' into which the active thread system is inserted at an angle of about 90 degrees. You should be careful that you pull all the bands as equally as possible, so that you obtain nice, straight outside edges.

By changing the positions of the coloured bands, a diamond-like pattern is formed. If you join the finished,

Cases

Braiding pattern

Y Y Y W B B B B

compact plaited band in a spiral to make a tube, if you have selected the correct diameter, this pattern will continue across the surface in the form of a grid. It is best to start in the middle of the strip by temporarily joining the bands with some sticking tape and then working outwards in both directions. The ends of the bands should be rounded off and the spiral structure continued until the strips automatically close in on themselves, when the final narrow band sides meet the outer edge. The end of the tube does not match the end of the plaited strip, but is about 7cm (3in.) away from it. For the case in the picture, the pattern concurs with a diameter of about 6cm. A 38cm (15in.)-long case is made from the 125cm (48in.)-long strip.

◀ A wide braided band is made in grey, black and white.
▼ The braided band is joined together in a spiral shape.

You can insert the zip using a sewing machine.

Depending on how closely you weave, these details may differ.

If the case is provisionally joined together, you can set the position of the zip and pin it up. One end is put on to the end of the outer edges, which are about 7cm (3in.) away from the end of the braided band. If you open the zip after you have pinned it up, you can easily sew it on with a sewing machine. You only need to set the outermost edges by hand. It is best to use a transparent sewing silk, so that the pattern of the weave is as little interrupted as possible. The zip length needed is 130cm (50in.) and this corresponds to the length of the braided strip and its width.

The black zip emphasises the spiral-shaped pattern and makes the case extra stiff. When you are using it, you can just unzip the upper section to reach what is inside. The case can, however, also be completely opened up again into a two-dimensional strip at any time, lending the object a playful character.

The second case in the picture consists of a wider braided band. A total of twelve, woven paper bands which have been halved in the middle were put into the following sequence at the start: two red, five grey, five black. This gives a width of 6.5cm (2 3/4in.). Because the dominant red stripes are positioned on the edge, not in the middle like the white ones above, it is not split into a diamond shape, but runs through the strip in a zig zag pattern. The three-dimensional joins of the bands give it a more spacious aspect.

This case is made according to the same principle as the above brush case, only the measurements are different. You need a strip 150cm (54in.)-long and 6.5cm (2 3/4in.)-wide, requiring a 157cm (55in.)-long zip. If you join the red zig zag pattern to the corresponding strip, the beginning of the zip should lie 16cm (6 1/2in.) from the end of the woven band, producing an approximately 30cm (12in.)-high, 18cm (7in.)-wide and 10cm (4in.)-deep, exquisite little bag.

Circular braiding

Circular braids have a long tradition in different cultures such as Japan or South America. This technique produces a compact, circular structure. Threads are crossed over one another and run together into a central area. Depending on which material and colours you use, how you organised the colours in the circle and according to which principle they exchange positions, you can produce very beautiful, highly imaginative patterns. In some regions circular braids were made without using any technical tools, directly on the hand. A braiding stool was used in Japan, known as *Maru Dai*. For a hand-held alternative, you can work on a simple braiding disc.

Paper yarn is good for circular braiding. It produces stiff cords with a nice structure which are suitable for making jewellery. Because any difference in tension is very apparent in this unelastic material, you need a certain level of practice to be able to produce a regular structure. It is therefore recommended that you try the technique out first on a softer material.

Jewellery (designed by Mäti Müller, Switzerland)
Materials needed:
- Corrugated cardboard (on both sides)
- Set square, pencil, scissors
- Paper string, strength Mn 0.8 (about 1m in red and black per hoop)
- Black sewing yarn
- Paper string, strength Mn 0.165 (about 20cm)
- Spray bottle

The eight-ended circular weave/plait for the bracelets shown here was made on a weaving disc. To prepare this, cut a circle out of a piece of double-sided corrugated cardboard. It should have a diameter of 10cm (4in.). Then divide it into 24 segments. At each of the marked points, cut 1cm (2/5in.) deep into the circle and make a hole in the middle using the end of the scissors.

Two red and two black paper strings (strength Mn 0.8) are used as weaving material here. Their length should correspond to two and half times the circumference of your fist, so approximately 55 to 75cm (20 to 30in.). The threads are tied into bundles, folded once and fastened to a skin in the middle with sewing silk (see the sketch on the next page). This tube is now pulled through the hole in the middle of the disc with the help of a feeder. The loose thread ends are spread around the circle in the shape of a cross. You need to make sure that the dark threads run vertically and the red threads horizontally to the direction of the edge of the disc where they are clamped to two adjacent clefts (see illustration on this page).

To begin braiding, hold the auxiliary thread tightly with your left hand underneath the disc and clamp it between your fist and thumb. Using your right hand, change the positions of the threads on the upper side. Let go of the black thread underneath on the left and latch it onto the slit directly next to the upper, dark thread pair. Now let go of the black thread on the upper right and latch it underneath on the right next to the now-alone black thread. There should be a black thread pair on each of the opposite sides. The disc is turned 90 degrees clockwise and the same thread change takes place as before, this time with the red threads. If you continue to turn the disc and change the position of the threads according to the same principle, a weave with a spiral pattern is gradually formed in the centre. Pull it down through the hole with your left hand and push the threads into the correct position on the surface so that you get the required tension and regular structure on the surface. If you are so far advanced with the weaving that the threads are no longer long enough to clamp, then pull the weave downwards out of the disc, cut the ends to the desired bracelet length with diagonally running edges, lay them across each other in the shape of a bangle and fix this point temporarily with a thin, strong sewing silk. Finally, tidy this up by with a parallel knot (1). For this you need about 20cm (8in.) of the thinner, black paper string (Mn 1.65). Lay a loop parallel to the weave across the connecting point. Wrap the long end around the stitched area several times so that about 3cm (1 1/4in.) are

Jewellery

A simple weaving disc for an **eight-ended** circular weave.

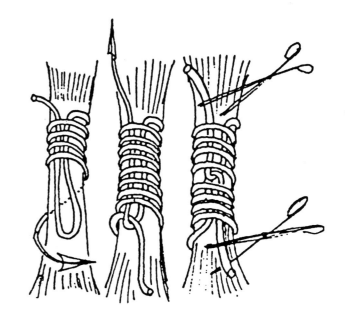

covered. Pull the end through the loop and set it by tightening the other thread end. The loop, and all the threads pushed through it, is pulled and clamped. The thread ends just need to be snipped off now and you have a clean, long-lasting border.

There are many possibilities for experimenting with circular weaves of this type depending on which yarn colours you combine, how you arrange these on the disc and according to what principle you swap their positions. You can also work with more threads, for example with 16 ends. In this instance, it is recommended that you no longer use the disc, but weave directly on your hand so that you can hold the tension more easily and can moisten the threads if need be. Other simple possibilities for the eight-end weave are shown in the adjacent sketch (2).

Parallel knots (1)

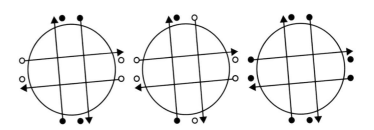

Variations of the figure-of-eight knot(2)

The future

Baskets

To make these baskets, a thick, natural-coloured paper yarn in strength Mn 0.32/4 was woven into a circularly arranged wooden bar framework. The framework is 25cm (10in.)-high and has a diameter of 20cm (8in.). This technique is a variant of partial weaving, using which you can make interesting, diagonal structures by wrapping the bars in a spiral. Depending on how many times you wrap around, whether, for example, you weave anti-clockwise or whether the loops are pulled up or down over the bars, there are many combinations and patterns which can be produced. In the background, you can see a big, plain-woven basket with a wire framework. This was made from the same material.

Baskets, Christina Leitner

Black jewellery

These pieces of jewellery made from black paper yarn, strength Mn 0.8 are by Lis Surbeck, Switzerland. They were made using a 7-loop weave and with no technical tools. In this primary weaving technique, a certain number of threads are knotted into loops and the knots are set together. The weave is made by continually pushing the loops into one another and crossing them according to a defined system. Because the tension of all the loops is not slackened during the weaving process, the number you can have is limited to the number of available fingers you have.

Lis Surbeck worked the weave when wet which is why it is particularly close and regular. To shape the piece of jewellery, she painted it with a mixture of white glue and water and wrapped it around a glass or a plastic tube with a suitable diameter. After drying, the weave remained in the spiral-like form, but was still elastic and flexible. Putting glue on the surface made it more resistant and gave the piece of jewellery a particularly stylish shine.

Black jewellery, Lis Surbeck

Stool

These brightly-coloured children's stools are made in the
Seestern basket and stool weaving factory in Männedorf,
Switzerland. They consist of a black, varnished wooden
framework and a simple covering, known as the Ticino
weave. In this widespread stool weaving technique, straw
is traditionally used. Instead of straw, a thick,spun paper
yarn (strength Mn 0.32/4) has been used here. By
interlocking the thick, taut strings crossways, a compact
surface is formed. At the end this is waxed to make it more
shower-proof. The material is very long-lasting and allows
very intense colours which cannot be achieved using the
traditional weaving materials. This is what makes the stool
so unusual and lively.

Seestern stool, basket and chair weaving factory

Covering a rocking chair

The original weave of this old Thonet rocking chair was
damaged and had to be changed. Natural-coloured paper
string, strength Mn 0.8, was used.

To achieve optimal tension, the weave is made on a
separate frame and only later fixed onto pre-drilled holes
on the stool and pulled tightly. On the framework of the
covering, there were nails above which the paper strings
were hung vertically and horizontally. The traditional multi-
directional weave comes about by crossing through this
grid diagonally in both directions with additional thread
systems. The plain-woven threads are pulled together in
pairs and diamond-like openings are formed which look
like small circles from the distance. This classic chair
covering is very stable but not very elastic. Traditionally, it
was made from cane, but for private restoration paper
string is a good alternative because it is easier to get hold
of than the natural material.

Covering of a rocking chair Christina Leitner

Design object

This detailed photograph is of design object which was made in several separate parts using the spring technique. It consists of narrow, vertically-running stripes made from thin, brightly-coloured hemp yarn. It is produced by crossing over the taut threads. After each row, compact paper strings are placed over the entire width, connecting the individual strips and producing broken points. The free-standing paper string ends on the edge can freely unfold and form a contrast to the strongly structured area on the inside.

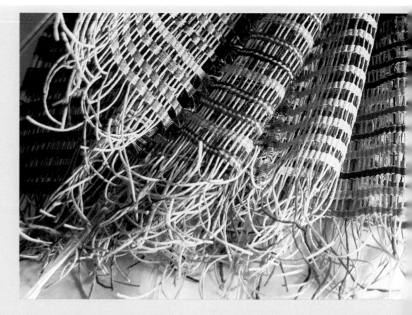

Design object, Christina Leitner

Little baskets

The Swiss basket weaver, Monika Künti, sometimes uses paper string in her work to bring out the colour. Sometimes she combines the yarn with classic basket weaving materials like rush or willow, sometimes she makes products from 100% paper. These small, spiral-shaped baskets are made of closely-woven paper string. The pretty effect on the border edge was achieved by twisting the woven ends up, producing tangled, curl-like bristles.

Little baskets, Monika Künti

Woven boxes

The lids of these boxes are woven from brightly-coloured strips of Nepalese loktha paper. By combining various colours in a set rhythm a decorative pattern is achieved. The upper part of the lid was freely woven at the start, then pulled diagonally over a handle shape, and turned up at the edges. The walls of the box were woven from the strips which met automatically at the sides. The ends of the strips are wrapped inside and the weave is reinforced with a mixture of wood glue and wallpaper paste. This produces robust and decorative lids for the square, thick, cardboard boxes.

Woven boxes, Christina Leitner

Folding stool

The basic framework of this folding stool was made of iron tubes and covered with paper string, strength Mn 1.65, using the sprung technique. The white and black yarns were put on a frame in parallel to the mounted, unscrewed side pipes and twisted into each other to produce the diagonal pattern which is typical of this method. They harmonise well with the diagonal elements of the structure. To stabilise the seat further, an additional cross thread was worked in in the same material after every row. This means that the structure is no longer elastic, but has a corpulent character and a very clear, almost graphic surface structure.

Folding stool, Christina Leitner

3.3 Weaving and decorating

Weaving means that a second thread system, the weft, is inserted into a taut warp at a 90-degree angle. Either mechanically or automatically, the warp is divided into two warp sections (sheds)usually using a loom. These sheds make it considerably easier to insert the weft. Depending on which principle the thread systems meet up, different weaves are produced. The simplest form is the plain weave in which a warp and a weft thread rise alternately. Over the course of cultural history highly sophisticated weaves have developed. In some of these warp and weft systems were combined using the most complicated weaves. In the majority of cases, the weft is active in weaving. It is woven into a passive warp. But there are exceptions: for example, the board weaving technique in which only the warp is the element which forms the pattern.

In the spinning process (as with weaving, knitting or crocheting) a textile structure is only produced whilst it is being made. Decoration techniques on the other hand, are characterised by the fact that an already available material is to be decorated with additional patterns afterwards. Embroidery, appliqué, mola production, and all colour and printing techniques come under this heading.

1 Sirpa Lutz, Switzerland
2 Claudia Bernold, Switzerland
3 Dorothea Rosenstock, Switzerland
4 Sanja Hanemann, Austria
5 Asa Haagren, Sweden
6 Christina Leitner
7 Christina Leitner
8 Christina Leitner
9 Claudia Bernold, Switzerland

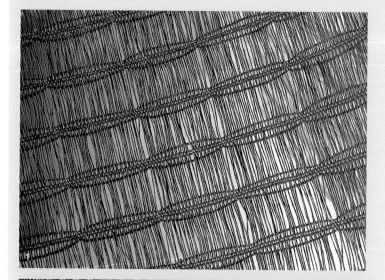

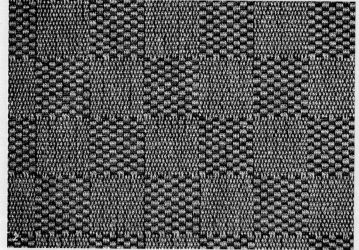

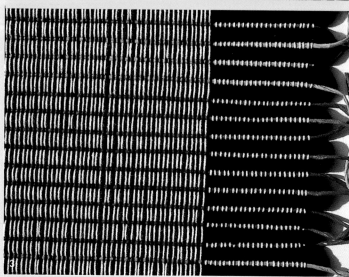

Board weaving

Board weaving is a simple, but brilliant technique, by which wonderfully ornamental bands can be made with only a few technical tools. Only the warp forms the pattern, as a rule, and is moved through the so-called board. The board usually consists of four square little cards which have a hole punched in each of the four corners and a warp thread runs through each of these. In the taut warp, all the boards lie parallel, close to one another and automatically form a weaving shed via the position of the holes. If you turn the entire board 90 degrees, the threads change position, producing a change of shed. The weft holds the spun warp threads together and forms a close, hard-wearing weave.

The pretty diagonal structure of this technique is really brought out when you use relatively fine paper yarns. The method is not suitable for shifu yarn.

Bracelet
Materials required:
- White and black dyed paper yarn in strength Mn 7.5
- Little square cards made from thin, but strong cardboard (cut to size playing cards are ideal), measuring about 6x6cm (2 1/2 x 2 1/2in.)
- Punch, writing material, Connective threads
- Twill band, about 1.5m (58in.) long
- 2 screw clamps

This bracelet collection (on page 114) was made from various board weaves. Quite a few preparatory steps are required for this technique, which makes the weaving itself all the quicker. The number of boards needed is determined by the amount of warp threads and therefore by the width of the weave.

    Each of the little numbered cards has a hole punched in each of its four corners. These holes are designated A, B, C, and D working clockwise from the top left. Later, a warp thread will be pulled through each hole, so you need four warp threads per card. To make the one in the picture, 8 cards were used, so 32 pieces of thread needed to be cut.

The first line gives the number of the card where you should insert the threads.

The second line tells you whether you should thread the holes on the board from the front or from the back.
/ = thread from front   \ = thread from back

	1	2	3	4	5	6	7	8
	/	/	/	/	\	\	\	\
A	×	×	×	×	×	×	×	×
B	×	×	×	●	●	×	×	×
C	×	×	●	●	●	●	×	×
D	×	●	●	●	●	●	●	×

The first column gives the 4 holes on the card, where A is top left and so on in a clockwise direction.

Each symbol in the point paper design stands for a specific yarn colour. Every hole on every card is defined on the grid system of this pattern and can be drawn up in the colour of the respective symbol. Here the marked yarn colour forms a triangle on the coloured background.

This grid tells you how many warp threads need to be cut per colour: here, it is 20 white threads for the background (x) and 12 black ones for the triangular shape (...). Once you have wound these 32 threads across two screw clamps at a distance of about one metre (1 yard), and cut through them, you can read in the first column of the point paper design the thread colours required for the four holes of the first little card and you can then pull the corresponding threads out of the warp thread bundle. Now take card number 1 in your left hand with the writing face up, thread the four ends from above through the holes and knot them together on the back. The card is then placed diagonally on the corner of a table so that the threads running through holes A and D hang down one side of the table and the others hang down the adjacent table edge. The knot should extend about 20cm (8in.) beyond the upper edge. The point paper design says tells you to pull the same colour through all the holes on the first card, but on the second card it is different. Three white and one black thread are pulled from

the warp bundle and the threads are put through the holes according to the colour distribution shown in the design. Once you have finished threading the second card, lay it directly on top of the first. Continue threading card by card according to this principle, but you must remember that after the fifth card, the threading direction changes. This means that you no longer pull the thread ends through the holes from above and knot them at the back, you bring them through the holes from behind up to the surface and join them together on the side with writing on. The cards which have been threaded from behind should also be put on the pile with the other four cards, still with the writing face up.

Once you have threaded the whole chain, string all eight cards together as a pack across all four side edges. Pull the upper knots equally tightly and secure them with a strong binding thread. Cut a loop from the ends and then hang this from one of the screw clamps. Now you can change the tension by pulling the tied set of cards down away from the screw clamp. This causes it to arrange itself in an upper (A, D) and lower (B, C) section and all the threads are tightened equally. You need to work very sensitively in this section, continually smoothing twisted threads underneath the cards so that they can be pushed further. Once the cards have been pushed down to about 10cm (4in.) before the end and both sections of the warp have been equally tightened, you can tie the lower thread ends together and push the board up a little.

Before you can begin weaving, you still need to make a bundle of weft threads. The same yarn is used as for the warp, usually the same colour as the outer warp thread. A contrasting colour gives an extra effect on the edges. If you wind the yarn several times around your thumb and index finger in a figure of eight movement and then set the centre by the end thread round several times, you get a small bundle which unwinds from the inside without losing its order when you pull on the beginning of the thread. This little bundle is sufficient for the width of the weave.

The tightness of the warp is produced by winding a

▲ A shed is made when you turn the board .
▼ By regularly turning the board, pretty patterns are revealed

Bangles

twill band around the lower knot and fastening this around your waist. You can therefore regulate the tension of the warp with your own body. Now you can finally begin weaving. To do this, undo the cards which are now upright in parallel (A and B on the top, B and C on the bottom) and insert the first weft. If you now turn the whole pack of cards 90 degrees forwards, maintaining the tension, then the holes change their positions. A moves down, C moves up so that from this simple movement a new shed from (A, B) and (C, D) is formed. The weft can than be inserted into this again. By turning again, more, different holes come to the surface, producing the pattern of the paper section design line by line. The weft is not visible in the weave and only appears in the form of the little dots on the outer edges.

If, as in the band in the picture, the cards are always turned forwards, a string of nothing but adjacent triangles is produced. But you can also change direction and this will turn the pattern upside down. By carefully combining forward and backward turns of the cards in a set rhythm, you can produce many different ornaments from one warp.

Thinner paper string (strength Mn 7.5 to 0.16) works well in the board weaving technique and produces wonderful, clearly marked structures. You can make a special feature using a paper string weft which is barely imaginable with any other material. Bcause the material is so stiff, you can insert it in the shed when it is too long, which gives decorative, stable little teeth on the outer edges. If you regulate the lengths differently, you can produce wavy borders. It is also possible to push the warp threads apart in certain places along the stable weft elements, so that the weft becomes visible and produces a pretty design. Whilst every other material would simply hang through and lose its stability, these plastic effects can easily be achieved with paper strings.

About 22cm (9in.) of weaving band are needed for a bracelet. Cut the strip up, bend it to form a bangle-shape and set the border provisionally with sticking tape, by letting about 1cm (1/4in.) of the weave overlap. The connecting point is tidied up by winding a black paper string around it and tieing a parallel knot (see page 105).

Because of the natural stability of the material, the bangles show a high level of inner tension and a perfect shape. The possibilities of the finely depicted, graphic structures are almost inexhaustible and several bangles, worn together, give a stylish, classic effect.

Here are a few more paper set designs for the bangles shown in the picture:

Oval shapes

	1	2	3	4	5	6	7	8	9	10
	/	/	/	/	/	\	\	\	\	\
A	●	●	●	●	●	●	●	●	●	●
B	●	●	●	×	×	×	×	●	●	●
C	●	●	×	×	×	×	×	×	●	●
D	●	×	×	×	●	●	×	×	×	●

By changing the twist direction, you can produce oval shapes.

Diagonal pattern

	1	2	3	4	5	6	7	8	9	10
	/	/	/	/	/	\	\	\	\	\
A	×	●	●	×	×	×	×	●	●	×
B	×	●	×	×	●	●	×	×	●	×
C	×	×	×	●	●	●	●	×	×	×
D	×	×	●	●	×	×	●	●	×	×

By changing the twist direction, you can produce diamond shapes.

Dot pattern

	1	2	3	4	5	6	7	8
	/	/	/	/	\	\	\	\
A	●	●	●	●	●	●	●	●
B	●	●	●	●	●	●	●	●
C	●	●	●	×	×	●	●	●
D	●	●	●	×	×	●	●	●

Long strips

	1	2	3	4	5	6	7	8
	/	/	/	/	\	\	\	\
A	●	×	●	×	×	●	×	●
B	●	×	●	×	×	●	×	●
C	●	×	●	×	×	●	×	●
D	●	×	●	×	×	●	×	●

Sometimes only with 4 cards, well-suited to separating warp onto länger wefts (optic sample)

Horizontal lines

	1	2	3	4	5	6	7	8
	/	\	/	\	/	\	/	\
A	●	●	●	●	●	●	●	●
B	×	×	×	×	×	×	×	×
C	●	●	●	●	●	●	●	●
D	×	×	×	×	×	×	×	×

By changing the threading direction, a 'knitted pattern effect' is produced.

Weaving

There are numerous possibilities for using a loom to make different kinds of paper strings into exciting textiles. Shifu yarn, as well as all strengths of European paper strings, can be used for the weft. Thicker strengths should sometimes be used damp. Shifu yarn is not recommended for use in the warp unless you are extremely experienced and careful with your work. European paper yarns can, however, be used for the warp without problems if handled properly.

When you are cutting a warp from paper yarn, it is important that you work very precisely because irregularities in the tension whilst weaving do not even out in time as they do with other materials, they increase. When cutting, it is best to put the spools of yarn into a nylon stocking to prevent them springing up, and bind the warp tightly several times before you use the shears. If the warp material struggles agains the warp beam, it is best to stick it down with some sticky tape before releasing the tension on the beam, so that the threads cannot spring back. You do not need to use a threading hook for threading because the yarn is so stiff and can then be pulled through the eyelets without any tools. The threads must then be set immediately as they would slip out otherwise. Once you have set up a loom properly with a paper warp, it usually runs without any problems.

Shawl (instructions from an idea by Lis Surbeck, Switzerland)
Materials needed:
- White *kozo* paper
- Something to rest on when cutting, cutter, iron ruler, pencil, scissors
- Sewing machine, white sewing silk
- White cotton thread in strength 20/2
- Flat loom with at least 4 shafts, weaving accessories

This exquisite shawl is made entirely from cotton and is very soft and pleasant to wear. Paper was only used for effect and not as a twisted thread, but in a different way.

Scarf

To produce the special effect weft, cut some *kozo* paper into 1.5cm (3/5in.)-wide strips and divide them in zig zags into small, equilateral triangles. These bits are now hung together using the sewing machine. Use a white sewing yarn made from cotton or silk for the upper and lower thread and set your sewing machine to the largest possible straight stitch. The triangles should now be sewn from the middle of one side to the point opposite, not sewing the individual parts closely together, but leaving some 'slack' in between. Upper and lower threads become interlocked without further stitching, and they twist loosely. After about 3 to 5cm (1 to 2in.), the next triangle is inserted. If you guide the triangle with your left hand and pull the finished thread backwards, a decorative string is produced which you should wind on to an approximate weaving fork from time to time. About 70m (80 1/2 yards) of this weaving weft are required to make a 176cm (73in.)-long shawl.

Use a white cotton thread in strength 20/2 for the warp. At least 250cm (97in.) of warp should be cut per shawl, if you want to add 20cm (8in.)-long fringes at both ends of the 176cm (73in.)-long weave. The weave density requires 12 threads per centimetre, so for a shawl width of 30cm (12in.), 360 threads are needed.

Once the warp has been cut, you can start threading. Because the cotton weave is being worked in plain weave and because only every fourth warp thread rises in the effect weft, you should thread straight through four shafts. The threads are double-threaded through a 60s sheet, where the six side threads on each side come in to the first two, so that both of the edges are doubly reinforced. Treadle twice to tie them in simple plain weave, pressing again only lifts the first shaft.

When you have finished setting up the loom, you can begin to thread the weft. The white cotton thread 20/2 is used here as well. It is set up to give a weft density of about 12 threads per centimetre. After 3 to 4cm (1 1/4 to 1 3/4in.) of plain weaving, you can begin to thread the effect weft. Use the third pedal for this, lay the sewn together paper triangles (with the top first) into the compartment and do not set off too tightly. This squashes the paper together into thick burls. Because the freely sewn sections of the effect weft between the triangles only consist of loosely twisted sewing silk, the triangles are flexible. If a burl does not fit into its place in the rhyhm of the weave, or stops on the very edge, the triangle can be easily moved into the right position. You then need to weave at least another 8 wefts of cotton thread and thread the next effect weft. The ends are cut off each time and not woven because they are well set in the weave. Leave them hanging and, at the end, cut 1 to 2mm off.

The effect weft is produced using paper triangles and thread

Once you have reached a length of 176cm (73in.) and you have therefore threaded about 200 effect wefts, you can finally weave another 3 to 4cm (1 1/4 to 1 3/4in.) in pure cotton, leaving at least 20cm (8in.) of warp material free up to where the next weave starts. This will be used later to make the fringeing

After the weave has been cut from the loom and cleaned up, you can make the fringes by knotting together four adjacent warp threads on the edge of the weave. Hold the material down on the edge of a table and turn two adjacent bundles of four parallel in the same direction. When you release the tension, the two form cords and you get a pretty fringe. Tie the end and cut all the threads to the same length.This is quite time-consuming, but produces a very stylish border which suits the fluid character of the weave. A simple edge can be achieved more easily with a simple seam.

Lamps

Materials needed:

- White paper yarn strength Mn 7.5
- Black Elasto-Twist (elastic yarn made from a cotton/lycra mix)
- Flat loom with at least 4 shafts and 6 pedals, weaving accessories
- Sewing kit
- Elastic hard plastic board, 52x3cm (20 x1 1/4in.) (sold in DIY superstores)
- Light bulbs with stands and flexes

These cylindrical-shaped lampshades are made from one weave in which white paper yarn (strength Mn 7.5) was used for the warp and the weft. The warp is 30cm (12in.)-wide and has a thickness of 12 threads per centimeter. This means that 360 threads need to be cut, plus an extra four reinforcing threads for the edges.

For the lamp weave, the threads are fed straight through the four shafts and inserted double into a 60s sheet.The two threads on the far outside are each doubled. The shafts are tied with a total of six pedals, so that the four outside pedals only lift one of the shafts, i.e. a quarter of the threads. One of the middle pedals lifts shaft 1 and 3 and the other lifts shaft 3 and 4. This enables you to weave a classic plain weave.

The first lampshade consists of a simple plain weave with about six weft threads per centimetre. For the second one, the 2cm (5/6in.) plainly woven strips alternate with 3cm (1 1/4in.)-long, unwoven, floating warp thread sections. Only the two middle pedals are needed for both these weaves and they are woven with the same weft material as was used in the warp.

For the third lampshade, the warp threads are woven in strips again. About 1.5cm (3/4in.)-long, plain-woven strips from each of five wefts alternate with 2.5cm (1in.)-long woven sections in which the warp threads have been divided into two sections. The upper weave, in which shaft 1 and shaft 3 are raised alternately, forms a thin plain weave over 2.5cm (1in.). At the same time, the threads from shafts 2 and 4 are floating and are not woven. If you cut these floats in the middle after you have taken the weave off the loom, then exciting, bristle-like rows of fringeing are produced. Particularly if the weave is shaped into a cylindrical shape, they stick out slightly and produce a plastic effect. So that the bristles do not fall out after the cutting process, two black wefts of Elasto-Twist are woven in at the beginning and end of the plain-woven strips. This pulls together and sets the warp thread sections.

The 30cm (12in.)-high lampshades have a diameter of about 16cm (7in.), so the weaves need to be at least 52cm (20in.) long. The inner frame shape for the bulb is made from hard plastic board (30 by 52cm)(12 x 20in.). 1cm (2/5in.)-deep cuts are made along both of the long sides diagonally in alternate directions. The distance between the cuts should be 2cm (5/6in.) so that the teeth this makes on both side edges can be easily fitted together.

A simple stand with a lampbulb is positioned on the inside of this valve and the flex is led through one of the lower tooth-like openings. Now, finally, the weave can be put on the cylinder, by pulling it flat over the tube. Then turn the end of the weave round and sew the base shape together on the seam by hand using blind stitches.

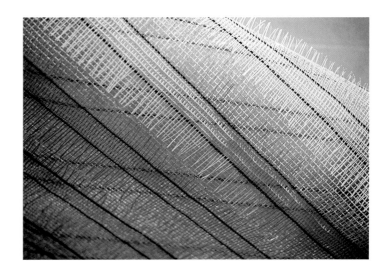

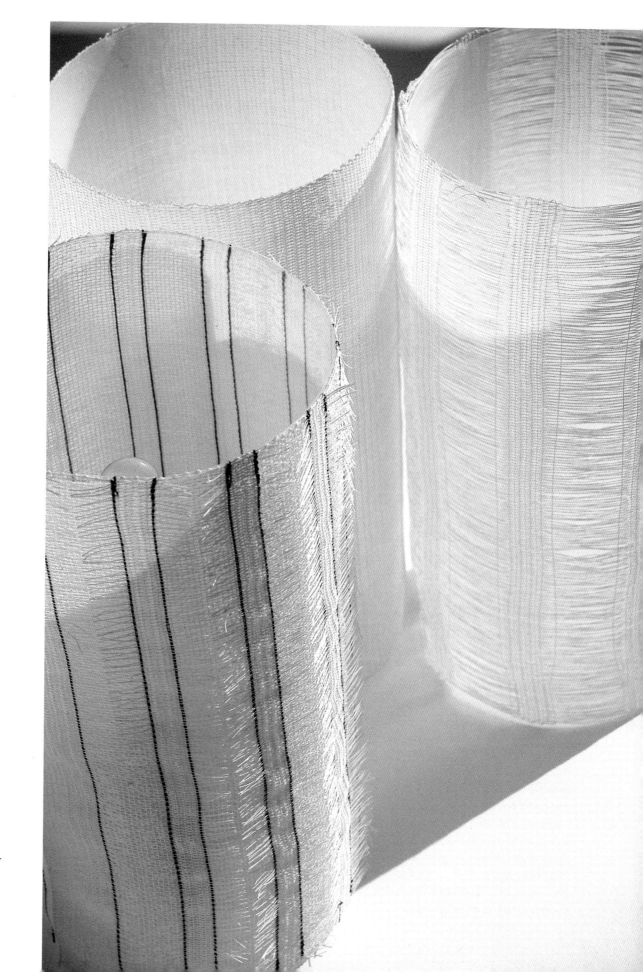

◀ The weave consists of paper
   yarn and Elasto-Twist.
▶ Lamps

Making pile

When you mention pile, people think of weaves or stitched structures where relatively shortthreadelements are inserted in such a way during the production of the material, that they stand away from the base material and thereby form a pile. These threads are not responsible for the inner cohesion of the material. These are additional decorative elements and the basic material could exist without them, theoretically. But its character is vastly changed by the pile.

Pile is most commonly used in carpets where short threads are inserted into the warp between the wefts. Depending on which knot is used, the length of the gauze and how densely the knots are placed, very different textiles can be produced.

This technique accommodates the stiff European paper yarn very well because it shows its characteristics off very well. However, there would be no point in using shifu yarns for this technique.

Design object

Materials requried:

- Brightly-coloured paper yarn in strength Mn 0.8
- High loom, weaving frame or simple stretcher measuring at least 75 x 30cm (30 x12in.) and circular bar of at least 25 lengths
- Weaving needle, weaving fork or large comb, scissors

This design object plays with the plasticity of the paper yarn and the radiating power of the colours. On the one hand, it can be purely decorative, on the other, if you place it in front of a window, it could function as modern form of isolation and produce interesting effects with different areas of incidences of light. It is a plain weave, because after every two wefts, a long piece of pile is tied in. This produces a very thick, powerful strip with a base of 13 by 65 cm (5 x 25in.), which equals a ratio of 1:5.

You can do small pieces of tieing work very well using a simple stretcher. Above a warp length of 80cm (29in.), the tension becomes so high from the weaving of the weft, that you need a device to slacken the warp. A high loom is ideal for larger dimensions, but can also be used for smaller pieces. Use the black paper string strength Mn 0.8 as warp material and stretch it in a density of three threads per centimetre across the frame. You need a total of 30 threads for the 13cm (5in.)-width, two additional threads reinforce the edges on the far outside.

The basic structure of the weave consists of a simple plain weave. Using a weaving needle, insert the black paper yarn strength Mn 0.8 in upwards and downwards movements into the warp again and weave row by row a

Coloured paper strings are knotted on to the black warp with Smyrna knots.

Two weft threads (closely pushed together) follow two rows of knots.

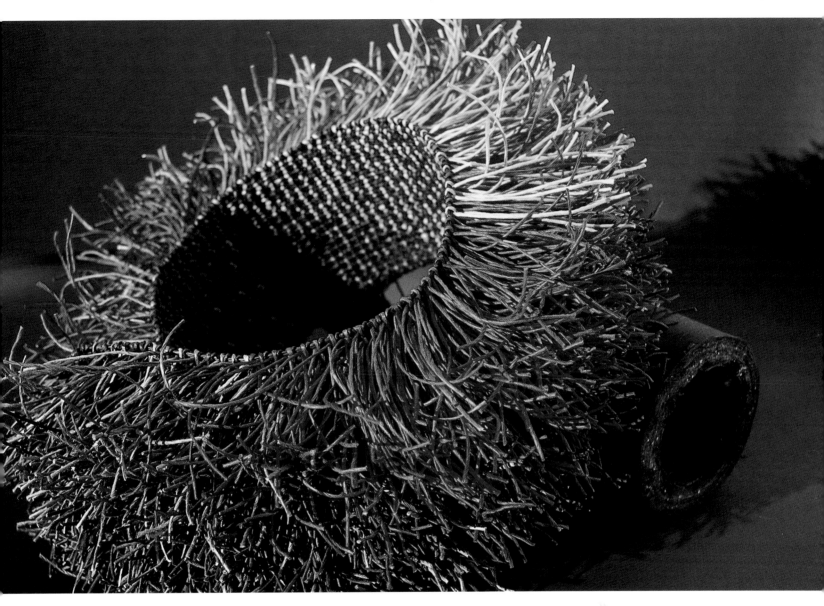

Decorative object

stable beginning strip. Be careful not to pull the weft too tightly so that the width is maintained. The best thing to do is to loosely insert it diagonally upwards into the warp and then push it down hard with a comb. To make it easier to form the sheds, you can tie in a braiding bar. Tie flexible loops of equal length around every second warp thread and thread them onto a wooden circular bar. If you pull this bar upwards, half of the warp threads rise and a shed is made. You can work a second bar in for the next group of threads using the same principle, or you can thread the second weft through without using any additional tools.

Once you have woven about 2cm (5/6in.), you can begin to make the pile. Cut the coloured paper strings into 20cm (8in.)-long strips and tie them into the warp using the so-called Smyrna knot. This traditional carpet knot consists of looping symmetrically around two adjacent warp threads. Lay the pile thread across the warp, let both ends dangle down on the outer sides of the warp pair and bring both pile sections up through the two warp threads. To stabilise the knot, pull the loop tight. This produces two vertical pile threads about 10cm (4in.) long. If you put one of these Smyrna knots on every warp thread pair across the entire width, you will 19 knots and 38 pile threads per row. They are set by weaving another two weft threads in plain weave and the whole weave is pushed together as closely as possible. After a few more rows of knots, there are two more wefts, a row of knots, two wefts etc.

If you weave the weft material so closely that on 13cm (5in.), 19 rows of knots with the corresponding wefts between them can be worked in, on a total length of 65cm

(25in.), five square fields with 19 x 19 knots are produced. This forms a great basis for the pattern. In the one shown in the picture, the shape of the triangle is examined. It is formed from the diagonals of these squares. They are each coloured differently. Inside one triangle, two to five related colours of yarn are combined to give pretty blended effects and so that the transition between individual areas of colour can be determined exactly. You can clearly see here, how similar yarn colours have a completely different effect in different surroundings. Whilst the colour groups on the surface merge seamlessly into one another across the length of the pile and clear borders become blurred, on the back, you can see the exact borders of the triangular shapes and the diagonal stripes. This contrast between the black warp and weft material produces a particularly pretty structure on the underside.

Once you have tied the whole pile, to finish off, weave another 2cm (5/6in.) of black weft in, cut the warp threads from the frame and clean up the border edges by setting the cutting places with wood glue and folding the woven edge strips inwards and sewing them up. This produces a heavy strip with a marvellous spatial effect. Because of the closeness of the weave, the base is very stable and can be easily bent to make circle or bow shapes. The image is always new depending on its arrangement or the angle from which you look at it. When you are planning an object like this, you should take into account that the material is quite expensive and that you need a lot of stamina for all the knots.

It is worth the effort though, because this pile weave has a very special character which probably cannot be achieved with any other material.

Embroidery and appliqué

If you embroider thin paper threads onto a base material, you can produce attractive, graphic structures. Because of their stiffness, thicker threads can also easily be appliquéd with crossover or zig zag stitches on the sewing machine. If they are closely strung together, you can produced very vivid surfaces. You can also achieve experimental effects if you put the robust threads onto transparent fabric or elastic materials when they are stretched. The decorative possibilities are almost unlimited.

Runner

Materials needed:

- Black linen material in plain weave (about 10 warp threads and 14 weft threads per cm) about 150x30cm (58 x12in.) in size
- White paper yarn, strength Mn 7.5
- Embroidery needle
- Black silk lining

This elegant runner is a linen weave on which effect lines have been embroidered with thin white paper yarn. The length of the strip is sub-divided into five, loosely embroidered squares which are each separated by a closely-woven, 2cm (5/6in.)-wide strip.

To prevent fraying, it is advisable to trim the material with rough stitches before you begin embroidering. The base material has a thread density of 14 threads per centimetre^2 and is woven in plain weave. To begin the first closely-woven strip, on the right lower corner of the strip about xxcm from the opposite side and 2cm (5/6in.) from the long side, thread into the material and embroider from right to left nine warp threads, up and down parallel to the lower edge. The effect thread is thereby inserted into the plain weave. Then, leave the yarn to run free across twenty warp threads, go into the weave along the weft again and pull the thread through the material again across nine warp threads. The embroidered lines of 0.9cm in length and the 2cm (5/6in.)-long floats alternate across the entire width of the weave.

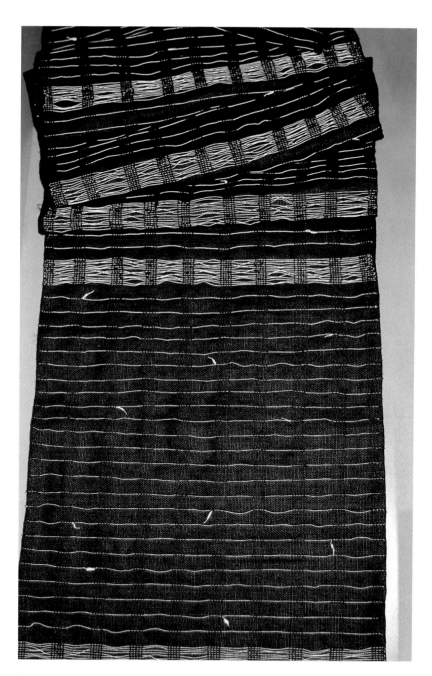

Runner

Once you have finished the first row in this rhythm, turn the weave round and embroider the second row along the next-but-one weft in exact parallel. After a total of 14 lines, a height of 2cm (5/6in.) is reached and small, stitched together squares (separated by the embroidered paths) are produced from the floats. If the embroidery threads run out part way through, knot the new one tightly onto the back of the weave and continue working. You can produce another effect, if all the floats do not run in parallel, but cross over the free threads. If you let the embroidery stitch run above the last line for up to four wefts in some places, and you fill the skipped areas later with other threads from above, the levels are changed and wave-like lines are produced. Whilst the short, embroidered point lines are all exactly parallel, the paper threads cut across one another in the floats producing graphic squares which are all slightly different.

In the large, loosely embroidered squares, the rhythm of alternating embroidered and floating areas is maintained. Only the distances between the lines are changed. Now, after every fourteenth weft thread on the basic weave, a paper yarn is threaded in, which equates to a gap of 1cm (2/5in.). If the embroidery material runs out in this weak area, do not knot the end, pull it to the surface, then cut it leaving 2cm (5/6in.). Then twist the thread end into a flame-like border to set it. If every second row is moved irregularly and a small, white eye-catcher of this kind worked in, a nice effect is produced because it loosens up the rigid surface a little.

After 26cm (10in.) and 24 embroidered rows, you should have a square shape and another closely-woven strip follows. A total of five loosely-embroidered squares and six stripes fill the entire surface. The result is an elegant runner which is lined with black silk and ironed at the end.

Small handbag

Materials needed:

- Stiff linen material (about 40 x 5cm (15 1/2 x 1 1/2in.))
- Paper yarn strength, Mn 0.8 (here in yellow, red, lilac and black)
- Transparent synthetic sewing silk
- Sewing machine, sewing needle
- Black silk lining
- Magnetic clasp

The decorative effect of this small handbag is achieved by appliquéing brightly-coloured paper strings onto a base material. First of all, the strong cotton weave is turned up about 2cm (5/6in.) on both sides, ironed out, and set with hemstitches. Now cut the paper yarns into 45cm (17 1/2in.)-long pieces and sew them with the sewing machine close together lengthways onto the base material. Use an invisible synthetic sewing silk for the upper and lower threads and set the individual strings with a zig zag stitch. The width of the stitch should be big enough so that the effect material from the sewing thread is enclosed, but not stick through. Because the paper string is so stiff, it is easy to guide and does not stretch. Sew the beginning and the end of each stitch and leave the surplus paper string ends sticking out about 2 to 5cm (5/6 to 1 1/2in.) from the edges.

The individual paper strings are sewn together with invisible synthetic sewing silk.

This turns the strictly guided lines into a random, playful mess at the end. The individual strings should be appliquéd so closely together that hardly any base material can be seen through them. An irregular colour sequence brings the surface to life and produces a strong, closely-woven structure.

Once you have sewn the entire width of the material apart from about two free centimetres (5/6in.) on both of the outside edges, fold the right corner parallel to the opposite side, turn the unworked edges inwards and sew the sides together by hand, using small stitches. You can now line the roughly 20cm by 25cm (8 by 10in.) little bag with black material and add the subtle magnetic clasp on the inside. Paper strings from the side walls work as a handle. These were not cut off, 80cm (31in.) was left loose and these can now be sewn to the opposite edge according to the pattern, subtly being inserted into the base surface.

Handbag

The future

Cushion covers

Swiss designer, Claudia Bernold wove this cushion collection from various different paper yarns. The cushions are 50 by 50cm (18 1/2 by 18 1/2in.) and done in various shades of grey or subtle blue tones. By cleverly combining the yarn strengths which are woven into a cotton warp, Claudia Bernold has produced a wide variety of graphic structural patterns emerging from the almost unlimited creative possibilities of plain weave. Closely-woven, corpulent surfaces are produced which impress because of their clear aesthetic value, their functionality and their perfect execution.

Cushion covers, Claudia Bernold

Shifu jacket

This traditional papier maché jacket, or *jimbei*, is the result of Mona Gustafssons' long and intensive examination of the Japanese shifu technique. In 1997, the Swedish weaver travelled to Japan for several weeks to learn the ancient method from the very beginning from a master. Since this time, she has never stopped working with this craft.

The traditional Japanese jacket was made from 26 handmade sheets of *kozo* paper. Cut into 6mm (1/4in.)-wide strips, one sheet produced about 60m of paper threads which could then be woven into a subtle, striped cotton warp in plain weave. The material is very compact, with a thread count of eight warp threads and twelve weft treads per cm^2. Mona Gustafsson needed at least 150 hours to make this hand-sewn jacket.

Shifu jacket, Mona Gustafsson

Table sets

These loose weaves are intended to be table sets or room dividers and were made by the Swiss designer, Schneider-Ludi. The mat-like character is produced by weaving the stiff paper strings into a partially threaded cotton warp. At intervals of 1cm (2/5in.) empty spaces alternate with closely-woven warp sections with 10 threads per cm, so that the paper yarn (strength Mn 0.8) comes to rest in stripes, producing a pretty, graphic structure. Because the weft material is so tightly clamped, the warp does not slip and it is possible to leave the paper strings loose at the woven edge of every row, and to then cut them off equally with a cutting machine at the end.

Table sets, Schneider-Ludi

Boxes

These solid boxes from 100% paper yarn were made by Eija Aro. The Finnish weaver has been working with paper strings from her home for a long time and prefers to use coarse material. The cubic boxes in the warp are made from paper yarn Mn 0.8. A weft in a matching colour made from thick paper yarn (strength Mn 32/4) is woven in to the warp. Because Eija Aro moistens the yarns before starting weaving, she can set off particularly strongly, thereby producing very thick, robust strips with an extremely high material density. To maintain the cubic shape, she wove strips which were three times as long as they were wide from this tough material and folded them into a U shape. Both of the free side sections were then filled with squares of the same material and the side edges were sewn together with large stitches.

Boxes, Eija Aro

Collection of bags

In the textile workshop of this Basel public hospital, about twenty patients, together with their four nurses, produce woven articles. This beautiful collection of bags is just a selection of what they make. Black paper string in strength Mn 0.8 is used for warp and weft and both systems are woven with a density of three threads per cm. Black, grid-like materials are produced which can be made into handbags, washbags, pencil cases, (glasses) cases and purses. Certain exposed areas are emphasised with colourful stripes, making the well-made, stylish bags even more vivid.

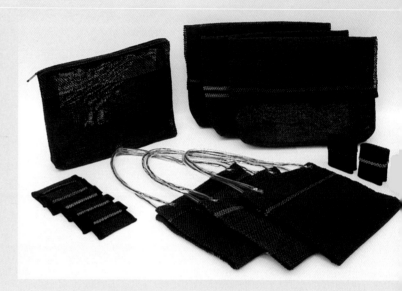

Collection of bags, Basel public hospital

Curtain material

Veronika Rauchenstein from Switzerland wove this wonderful, transparent curtain material from thin paper string in combination with polyester thread. In some areas, the warp is divided into two loose layers of material. Loose paper strings float between these. They can move freely around the square areas and display their stubborn character. This produces interesting overlaps and the multi-faceted weave hangs particularly well and lightly.

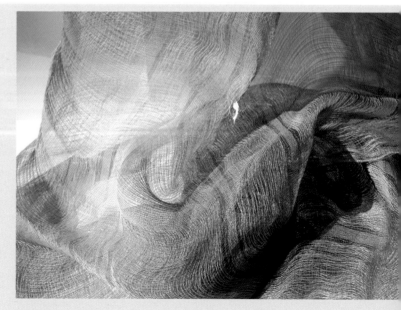

Curtain material, Veronika Rauchenstein

Silk paper scarf

The Swiss weavers Franziska Käser and Erika Wyss run the Webzettela workship in Bern together. In their workshop, they produced a collection of shawls in double weave. These consisted of one woven side of black silk and the other made from Nepalese shifu yarn. By swapping these over in sections, very effective weaves with a bark-like folded structure were produced. This effect is produced by laying the finished weave in warm water. This makes the paper thread go wavy and in turn, is transferred automatically to the second layer of weave via the technical binding points. This gives the material its unmistakable character.

'Vulcan music'

This detailed piece is from the 'Vulcan music' triptych tapestry by Renate Egger from Austria. To make the natural-coloured background, she mixed linen yarn with the Nepalese shifu thread to produce an interesting texture. The burled character of the material, combined with the slit technique of the Gobelins weaves, brings the large, self-coloured surfaces to life. The special qualities of paper yarn – its natural body and its resistance to pests – also fit in very well with this piece of art.

'Vulcan music', Renate Egger

Silk paper scarf, Franziska Käser und Erika Wyss

Design weave with little paper tiles

These large woven strips made from 100% paper are room
dividers. They consist of a warp of thin paper yarn,
strength Mn 7.5 and the weft material is even finer. The
warp has 16 threads per cm and sections of it were woven
hollow, and the spaces which this produced were filled
with square paper tiles in a random pattern before they
were closed. This padding material which indicates the
origins of the yarn, was dyed with woodstain in various
shades producing an interesting interplay between
transparency and density.

Design weave with little paper tiles, Christina Leitner

Lamella weave

Paper string worked into a warp made from silver wire
produces an interesting combination of material. To make
this material, the warp was woven in lines with 12 threads
per cm. A plain-woven block alternates with the sections, in
which the warp was divided into two parts, producing
separate, overlapping layers of material. After removing it
from the loom, one of the layers was cut open in the
middle and bent open irregularly. This produces a three-
dimensional wall object with wave-like patterning.

Lamella weave, Christina Leitner

Weave and latex experiments

These samples of material are from a black and white sample collection, where the weaving technique was combined with a latex application process. This produced materials with a warp made from white paper string in strength Mn 7.5, mainly woven with black Tussah silk in various weaves. The warp material was also put on to elastic jersey material with black latex in various arrangements. The thin, white paper thread with so much character is the link in this unusual combination of stiff material with soft, fluid latex material.

Weave and latex samples, Christina Leitner

4 Gallery

▲ Osamu Mita (Japan): Double weave in wool and shifu threads
◄ Ritva Puotila (Finland) 'Scent of lemon'

Today, there are artists and designers worldwide who use paper to make textiles. As a representative sample of the many exciting developments, you will meet twelve people in this chapter who all work with the textile side of paper. However, they have found very different, independent ways of expressing themselves through this medium. What they do have in common, though, is that they are devoted to 'their material'. Some of them have been plumbing the depths of possibility and testing its limits very seriously and with great passion for many years. Their work makes it clear how many different aspects paper textiles have to offer. It provides an ideal projection surface for many different cultures and working methods like no other material. From Japan to Finland, from Nepal to Switzerland, from the Philippines to America, Paper is essentially a 'descriptive material' and not just because you can write on it, but because it itself carries active potential within it.

In Japanese textile art, paper holds an important position in various forms today. But there are only a handful of people who still make shifu yarn. Some examine it strictly traditionally, and others work more experimentally with the ancient method. Since shifu was declared a 'national cultural heritage ', it is encouraged by the state. Nonetheless, the skills required to make it are in danger of dying out because very few young weavers are interested in this art form which brings in very little money. Outside of Japan, shifu was first made in the US. In the 1970s, there were textile experts who heard about the Japanese tradition and went on research trips to track down this culture and write about it. Shifu was very current with American weavers for several years before they lost interest again.

Although shifu is virtually insignificant in the US today, people are just beginning to become curious about it in Europe. There have been a few specialists in Switzerland for a couple of years now. They devote themselves to making paper material by hand, produce shifu themselves and gather information and contacts on Asian experts. The commitment of a few people has produced a small, dynamic scene which is now spreading to other countries. Making paper strings from wooden fibres is more widespread than shifu in Europe. The boom of European paper string started in Scandinavia in the 1980s. In Finland, people began to use the wartime material again to make innovative carpets and accessories and to develop the yarn quality further. Although the historical material was almost always natural brown, you can get it in many pretty colours today from several different yarn manufacturers. These aesthetic products from the North brought many designers' attention to the strings and in the meantime, the enthusiasm had to spread to the whole of the European textile scene. Paper in its various forms is still very current, particularly when used in objects for the home.

In the industrial sector, carpet companies in particular have increased their production of paper strings. Recently, well-known interior and fashion textile designers have also discovered paper for making textiles. Cut into strips, twisted or glued, it is made into the best, forward-looking, qualities of material using the most modern machines.

These experiments, produced in the last few years, clearly demonstrate how adaptable and constantly surprising paper yarns can be. From filigree, soft structures to robust, archaic carpets, almost opposing effects are possible with this material and its creative potential is nowhere near exhausted.

4.1 Japan

Sadako Sakurai

Sadako Sakurai is one of the few masters of Japanese shifu production still alive. She is probably the only one who continues to make wafer-thin threads from 2mm (1/12in.)-wide paper strips according to the old Samurai tradition. She then makes these into sophsticated materials. She has devoted herself to the fascinating transformation from washi to shifu for more than thirty years now. Her aim is to prevent the secrets linked to this process from fading into obscurity. She wants to provide the loud everyday Japanese life today with calm evidence of its flourishing past.

Sadako Sakurai lives with her husband in Hori-Machi, Ibaraki-ken, a small town to the north-east of Tokyo. Her simple house also serves as her workshop. The fact that her loom, spinning wheel and other accessories for making shifu are in her living room, gives us an idea of how central this craft is to her life. For her, life and work are one and the same. She owns little, but what she does have is high in quality. Her equipment has lasted for decades. It is very simple, but functions perfectly and bears witness to the high aesthetic value of usefulness.

The shifu weaver gets the paper for her threads from neighbouring paper makers who make very special sheets for her from kozo fibres. They measure 90 by 60cm (35 by 23 1/2in.) and are made in such a way that optimal tensile strength is guaranteed when they are cut. Mr Sakurai usually folds and cuts the paper. If he makes wafer-thin, 2mm (1/12in.)-wide strips, about 250m (270 yards), of thread can be obtained from one sheet. 50 sheets are needed to make 12.5km of yarn, roughly the amount needed to make a single Shifu kimono.

It is amazing how carefully Sadako Sakurai twists the cut sheets into almost perfect threads on a cement stone, without a single strip tearing. Using a simple Japanese spinning wheel without a foot pedal, the pre-rolled strips are twisted even more strongly after the tearing. Sadako Sakurai kneels on the floor to do this. She needs to turn the wheel by hand and turn the yarn on the awl and wind it on. Physically, this step is very tiring and requires a lot of concentration to produce an extremely thin, almost silk-like spun yarn.

Before the weaving begins, some of the yarns are dyed with plant dyes made according to traditional recipes. Sadako Sakurai grows most of the plants she needs in her own garden. Her favourite colour is indigo blue, but she also uses onions, walnuts and other plants. The dyeing process alone takes many days, but the resulting vivid, permanently-coloured areas of the weave are worth it.

The loom has twelve shafts and allows a weaving width of up to 120cm (46in.). You would think that a European would never be able to produce such precise weaves on such a piece of equipment. We are used to working on sturdy looms, which provide reliable support like a deeply-

rooted tree and which bear the power of casting on. Japanese looms, on the other hand, are light and filigree, they almost seem flexible. In a physical interaction between the weaver and the loom, a technically perfect weave is produced weft by weft.

Sadako Sakurai usually weaves the Shifu yarn into a cotton or silk warp. She produces traditional Japanese stripes and checked patterns in simple weaves, but also Ikat materials. This ancient, sophisticated process produces surprising coloured effects, by partially dyeing the yarn before weaving it in. The finished materials are then washed and dried in the sun several times so that they become soft and supple. They are then made into traditional kimonos and obis, also shirts, ties or textiles for interior furnishings. Work overalls are made from slightly coarser weaves. They are unexpectedly long-lasting.

Although this shifu master has already received many honours and has exhibited in numerous places, she can barely live from her craft. It is impossible to think of an adequate recompense for the effort and hours of work which go into her craft: to make a single shifu kimono requires two months of very intense work. Even if the prices of the products seem expensive, they bear no relation to the time spent making them. When you purchase materials from Sadako Sakurai, you are also buying a a piece of her idealism and commitment.

Sadako Sakurai has completely devoted herself to her craft. She seriously researched the roots of shifu in her own country and committed herself to the preservation and government support of this tradition. She often took in interested parties from abroad, and freely gave of her priceless knowledge. She played a considerable part in making shifu known to experts beyond the Japanese borders where it was then also practised. You realise when you are holding Sadako Sakurai's materials in your hands, that it is impossible to attain the perfection of the Japanese originals and that attempting to imitate it is virtually pointless. Visitors think they know all about the production process, but it remains a secret. People are almost reverently amazed how Japanese cultural history can be caught up in a few centimetres of material.

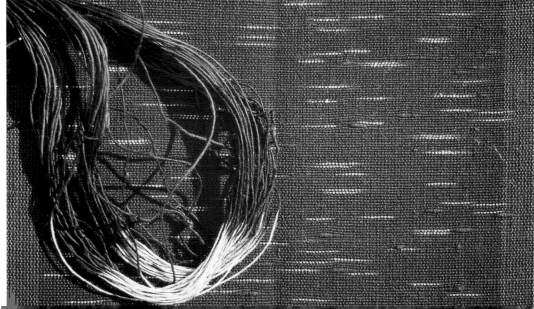

Naomi Kobayashi

Japanese textile artist, Naomi Kobayashi has been well-known internationally for her highly-aeshetic work for years and she has been awarded numerous prizes. Like Sadako Sakurai, she also uses handmade Japanese paper. However she does not examine the tradition of her country in the form of a passed-down craft, she transports its contents into an individual, contemporary language. You can still sense the centuries-old spirit of Japanese textile culture in her work.

Naomi Kobayashi (1945) studied art and textiles at the Masashino Art University in Tokyo. When she graduated, she began to work for a manufacturer in Kyoto, where she met her husband, Masakazu Kobayashi, who was also a passionate textile artist. Inspired by European examples, in the 1970s, the Kobayashis were amongst the pioneers who stopped using the textile material for wall tapestries, but used it to make three-dimensional installations for the first time. They marked the start of the beginning of Japanese textile art today.

The now-freelance Kobayashi couple live about an hour's drive north of Kyoto. Their workshop is in the

middle of nature and a small exhibition area has been built next door. Naomi Kobayashi does a lot of her work in the open air, because for her, nature is one of the most important sources of inspiration in her work. The lightness of a spider's web, the silver flash of a ray of light reflected on a disturbed water surface, the transparency of a light morning mist – she wants to fix these fleeting impressions in her work, because she is convinced that even the big, fundamental structures of life are reflected in these small things.

It became clear to Naomi Kobayashi from many experiments that paper is the material which best expresses her ideas of transcendency and lightness. Perviousness to light, brightness and last, but not least, the deep cultural and historical significance which has been ascribed to paper for centures in her country, made the material exciting to her. She therefore began to create filigree, geometric shapes from it. This is how she came to produce delicate suggestions of cubes, rings or tubes from reinforced paper threads and thin little paper tiles.

To achieve the effect of absolute weightlessness, Naomi Kobayashi developed her own technique for these objects. Her open sculptures are worked over a base shape. This consists entirely of small wooden blocks which are stacked on top of one another to give the whole shape. The artist covers with base with a thin layer of film and begins to wind paper threads (which she has spun) around it crossways to produce a grid-like structure. Once enough of the geometric shape is covered to be able to guess at the coat from a few lines, she stops and puts *konnyaku* on the paper threads. This natural substance hardens the threads and also makes them waterproof. In her next step, she lines some of a few selected areas between the lines (produced in the previous step) with more paper. Naomi Kobayashi uses the finest *gampi* paper for this. These let a lot of light through. She cuts them into triangles, wraps their edges around the paper threads of the sculpture which restrict the stretched areas and then sets them with konnyaku again.

When the paper is dry, Naomi Kaobayashi starts to work the base material from the inside out. Because the basic shape is only made of small wooden blocks, she can (after breaking through the thin layer of film) push individual blocks out through the remaining openings. This step requires a lot of sensitivity so that the delicate

wrapping is not damaged. An extremely thin skeleton is left behind, covering nothing but air. Even this seems almost too heavy for the fragility and brilliance of this object.

Naomi Kobayashi lets her shapes float in white spaces. They are camouflaged in such a neutral environment and are only noticed when you look twice. They express both the eternity and the fleeting nature of the moment. She loves circular arrangements because she sees a symbol of the cosmos and the cycle of nature in them and it is true, that when you look at her objects, you get the feeling that you are very close to these hidden truths.

In her most recent works, Naomi Kobayashi has returned to her original techique – weaving. Instead of cotton or silk however, she now works with handmade sheets of paper on which she writes calligraphic characters almost meditatively. When she writes, the black Japanese ink she uses penetrates the paper across the wide brush and can also be seen on the back of the paper. The pages, weighed down with meaning, are then cut into 2cm (1/2in.)-wide strips, moistened and twisted into compact pieces of thread by hand on a porous stone. This so-called *koyori* technique is thought to be the precursor to the shifu process. Whilst an 'endless yarn' is obtained from one piece of paper in shifu, the length of the thread is limited to the width of the sheet of paper in koyori. The short pieces of thread used to then be waterproofed and were mainly used for braiding and lacing work.

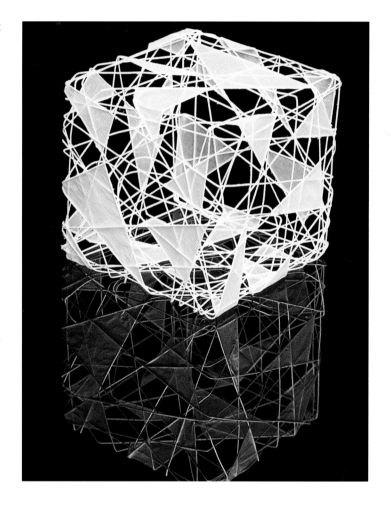

Naomi Kobayashi references this tradition in her work and also to the oldest preserved shifu dresses from the Edo period. These were made from old, used up accounts books and still display the hidden chracters in the form of dark speckles. The black traces of the ink are also visible in the twisted pieces of thread in Naomi Kobayashi's work. They are woven into a warp producing dark, thickened places, grey flecked sections and completely white areas. It is amazing how graphic the effects are and what an effect of depth is produced. When you look at it from a distance, some of the walls of the weave look like abstract black and white photos of moving water, wind and distance. This is also what the characters are hiding on the inside. Naomi Kobayashi is capable of formulating true poetry with threads instead of words.

Philippines

Asao Shimura

Asao Shimura (1950) is an internationally recognised Japanese papermaker and book artist. He has been working intensively with the the great paper tradition of his home for over 25 years now. He has been director of many workshops and has published a small series of mini books about the history of paper in various countries. He organised washi tours for years. He led papermakers and other interested parties from across the world to various renowned workshops throughout Japan to discover the traces of paper and open up this exciting culture to them. This provided him with many international contacts and meant he was invited to workshops across the world. Meeting Sadako Sakurai and other shifu weavers personally triggered his interest in the unusual qualities of the material and with great enthusiasm and seriousness, he began to examine different papers to see whether they could be used for shifu.

Asao Shimura became aware of pineapple fibres as a source material for making paper at a workshop in the Philippines. He was immediately impressed by their qualities. The long, robust, straw-like plants are called pina (from pineapple) and have been spun into yarn for more than 300 years, producing thin, relatively stiff weaves. However, people have only been using the fibres to make paper for a relatively short time.

Asao Shimura discovered that this material was perfect for Shifu papers. He has now been living in the Philippines since 1989 with his wife and four children. They live in a small, remote village, 1000m above sea level. There are large pineapple plants in this region. The fruit of which is mainly grown to sell on. In his workshop, the sinewy fibres of the leaves are made into a pulp from which extremely fine, tear-proof sheets are produced using the traditional Japanese papermaking technique.

In another area, about three hours drive away, in the province of Aklan, shifu is produced from paper from Asao Shimura's workshop. Five women have been working ambitiously for him on this project for more than five years. Originally, the materials were intended for a Japanese weaving mill, which pushed shifu projects and gave the Philippines assignments. Because of the current, bad economic situation in Japan, however, this cooperation was stalled. Now Asao Shimura suports the shifu weavers himself so that the craft and the knowledge gained from it do not disappear. He sees the Philippines as a forward-looking place for shifu because of its basic economic and ecological conditions and hopes that the tradition can survive here even though it is threatened with extinction in Japan.

The shifu materials which Asao Shimura developed in conjunction with his weavers is very impressive. Because the pineapple paper is soft, yet incredibly stable, entire, very thin strips of paper can be cut from it and twisted into the finest threads. The strip width is usually no more than two millimetres (1/12in.). The material is then woven into fine materials on simple bamboo looms. Because the distance between the material beam and the warp beam is very large, the warp is used as little as possible when the compartment is opened. This means that even very brittle, sensitive materials can be easily worked.

All the materials are woven in plain weave, but they are surprising because of the high thread density and their material quality. On the one hand, shifu yarns are used in combination with a cotton or silk warp which produces a pretty, firm material. On the other hand, very thin, lightly spun pineapple fibres are also used as warp material and up to 20 threads per centimetre are threaded. This produces very fine, airy weaves. But Asao Shimura's greatest passion is *moro-jifu*, material made from 100% paper. With great technical skill, the weavers work the paper threads, using 10 to 18 threads per centimetre, into the warp and the same number into the weft.

When the materials come off the loom, they are initially quite stiff. After they have been put in warm water ten times, then pulled in all directions and dried in the sun, they have a completely new texture. They become soft and very absorbent, but they also solidify into thick materials with a light, cooling character, slightly reminiscent of linen. For some materials, paper is used which has been dyed with plant dye in soft pink, grey or yellow tones. Asao Shimura also experiments with paper which he treats with konnyaku before making into yarn. This makes the sheets tougher and water-resistant. These qualities are then transferred to the material. It is more difficult to twist them into yarn, but because of their stiffness and the natural waterproofing, they have their own very special attraction.

These materials make it clear that the main intention of Asao Shimura's work is experimenting and systematically researching different materials within the framework of a great awareness of tradition and sound background knowledge. Selling his materials is less important to him than the fascination which transformation of different plant fibres into materials which can be worn brings to him.

Ecological, cultural and historical considerations play an important role for him in every step of the process. He has already introduced many interested people to this special fibre material and the possibilities it offers at pineapple workshops in the US, Denmark, Australia, France and South Africa. In this way, he has also made a major contribution to making shifu known outside Japan.

Asao Shimura is currently the only person producing shifu in the Philippines, but he is convinced that craft as a social hobby could soon spread. In the near future he wants to add fibres from banana plants, kozo and other native papermaking plants to the the strong character of pineapple fibres as a source material for making shifu. It is exciting to see how traditional Japanese paper weaving will be expanded in the Philippines.

4.3 Nepal

Deepak Shresta

If there is anything left in our globalised world which is still exotic, then a Nepalese man who makes shifu and then sells it to Europe, Japan and America, has got to be in the running. Deepak Raja Shresta from Kathmandu has realised his dream by doing this. The unconventional shifu maker studied for various careers, including architect and tour leader. As an interpreter, he took part in the International Paper conference in Hawaii in 1985 and interpreted for the Nepalese papermaker. He got his first small insight into the world of papermaking there. The Japanese shifu tradition was also discussed at this conference. This time-consuming, almost ritual process immediately entranced Deepak Shresta and he realised that there was a lot of similarity between the original Japanese papers and those from his home, Nepal. This is how he came up with the idea of attempting to make shifu from Nepalese papers. When he came back from the conference, he began his first experiments on an old flat loom as a second job. His first attempts seemed promising because the material was easy to twist and weave, and did not present too many problems, but still had a very original character, contrasting with that of the Japanese models. Technical weaving problems and uncertainty about the subject meant that Deepak Shresta hit his limits quickly time and time again. In 1988, he therefore went on a study trip to Japan to learn the original method from the very beginning from a traditional Shifu weaver and to see the historical and contemporary Shifu products for himself.

Back in Nepal, he founded his own shifu workshop in Kathmandu. Many members of his family now work there. Although it is very difficult to survive on the basis of producing this material, Deepak Shresta has made a name for himself amongst experts in the last 15 years with his special workshop. He is still the only person in Nepal producing materials of this kind. He gradually produced

something original, an unmistakable, new material quality whose source of inspiration may lie in Japanese tradition, but which shows a completely different aspect in its new environment.

Deepak Raja Shresta does not produce the paper to make shifu himself, he buys it from a small papermaking village, high up in the Himalayas, about 70km south of Mount Everest. Rustic paper from the native loktha plants are still made here according to the traditional pouring method which is widespread throughout Nepal. It is a long, arduous journey which Deepak Shresta undertakes again and again to reach his source of material. He carefully selects each sheet, making sure that they have been poured as equally as possible, that they have no large dirty areas or knots and that they weigh from 17 to 25g/m^2.

In Kathmandu, the papers are then folded by the Shresta family women. They then cut them into strips, roll them out on a basalt stone and finally, spin them into threads on old Nepalese spinning wheels. The whole workshop is very simply equipped. The women sit on the floor while they work and some work in the open air, others on the veranda. They have now established a routine and almost produce spools of thread in series. You can tell from their work that effective production is important for the family business and to ensure that they can live from it.

Deepak Shresta usually weaves the materials himself. Although he has no weaving training, he has learnt how to master the old machines well. He often weaves the Shifu yarn into a natural-coloured or black cotton warp and in combination with the paper threads, this produces a pleasant haptic quality. Because every sheet of paper he works has a slightly different colour tone and the strength of the thread varies slightly because of the different strengths of the sheets, the weaves never have a uniform suface. Hints of stripes, differently shaded areas and a varied texture really bring the material to life. They tell stories of the different areas where the plants grew, of the weather, papermaking methods, and people. Every piece is therefore preciously unique and a paper testament to the beauty of imperfection.

In addition to combining paper with cotton and silk, Deepak Shresta also tries to make *moro-jifu* material, where shifu yarn is used in both thread systems. He also combines the paper threads with other qualities in the weft and thereby produces rib-like materials (*kabai-ori*). In addition to these structural effects, he also works on the coloured pattern structure, by weaving coloured strips in, printing on finished materials and experimenting with plant dye. He sometimes also creates items of clothing from his own weaves, always trying to go another step further in his development. .

Deepak Shresta is now quite well-known on the paper and textile scene. For Nepalese people, his materials are exorbitant, because despite his low revenue, the products are extremely expensive because of the enormous effort which goes into making them. He therefore exports them, primarily to Japan and Europe and he has a very close connection with Switzerland. Fred Siegenthaler from the Basel paper mill visited him in Nepal in 1988 and then reported his work in a brochure, which incited interest in his material there. The Swiss fashion designer, Gisela Progin has obtained material and yarn for her own creations for years. She then sold these on to interested people in Europe. This produced a flourishing relationship with Switzerland and Deepak Shresta was invited to give a presentation on his skills at a workshop there in 1997. American lovers of shifu have also started to buy from him because his materials and clothing were recently displayed in a few galleries there and he gave some lectures about his work.

Deepak Shresta's enthusiasm for shifu is internationally understood and can be seen in his original, rustic

materials. They are simple, genuine and make an immediate impression on one because they reflect an entire country and its culture in a tangible and understandable aesthetic way.

Gisela Progin

Swiss textile designers' increased interested in the Japanese shifu technique in recent years is all down to Gisela Progin. This exceptional artist has also examined clothing and objects in her own way and she contributed, with her own passion and open commitment to shifu materials, to the material becoming well known beyond Switzerland and throughout Europe.

Gisela Progin (1957-2002) studied to become a kindergarten teacher, but she soon moved into art. When she stayed in America, she ended up in a fashion workshop and began to design clothes professionally. Her extravagant, brightly-coloured silk garments were very reminiscent of Japanese kimonos because of their reduced cuts.

Back in Switzerland, she took the certificate of tailoring examination and founded her own fashion workshop in Murten in 1984. From the very start, she specialised primarily in silk. Together with the artist Vernessa Riley-Foelix, who painted Gisela Progin's garments and thus gave them a note of individuality, she produced many exhibition projects in Switzerland, Germany and the US.

Gisela Progin came up with a completely new material quality for herself in 1993 which after a few drawings hit her 'like lightning' and led her in a completely new direction stylistically. When she visited the Basel paper mill, she came across a small piece of original shifu material in a brochure about Nepalese shifu weaver, Deepak Shresta. She was immediately fascinated by this unusual material and wanted to get the bottom of it. Because very little information about this exotic technique was available in Switzerland at that time, Gisela Progin decided on the spot to travel to Japan to research the material, with the dream of being able to use the special fabric for own creations one day.

Thanks to financial support from the Commision of Culture in Freiburg, she was able to stay in Japan for six months in 1993/94. She then received a further, widely effective contribution: before she left, she contacted textile makers throughout the whole of Switzerland and told them of her travel plans. If they offered to contribute financially, Gisela Progin would send them information about shifu at

regular intervals for the duration of her trip. This paved the way for the widespread enthusiasm for shifu.

But it was not as easy as she had originally assumed to find out about shifu in modern Japan. She worked in a fashion workshop at first, learnt how to make paper herself in a factory and finally spent a week with a couple, Mashiko and Tadao Endo, a papermaker and a weaver who had devoted themselves to maintaining the shifu method. She had arrived at the traditional roots. Filled with intense impressions, she returned home.

It soon became clear that the Japanese shifu materials were out of the question for her own clothes because the prices were completely beyond her reach. She nonetheless wanted some paper weaves, so Gisela contacted the Nepalese producer, Deepak Shresta because his shifu materials were cheaper. They had a completely different character to the Japanese models, their power of expression and the pleasant qualities for wearing them were immediately of interest to Gisela Progin. She developed simple, almost archaic-seeming clothes which almost all showed a characteristic, bark-like, pleated structure. Gisela Progin achieved this unique surface effect by leaving the Nepalese materials (which were woven with a lightly overspun paper thread in the weft) in warm water overnight. Like all paper when it gets wet, the paper material began to form waves. Because the cotton in the warp shrinks in warm water, it sets the thrown up paper threads by contracting. If you iron the material out when it is wet, the cotton warp moves and presses together to such an extent that the surface becomes smooth, but the shrinkage cannot be reversed. If you let the material dry without ironing it, the uncontrolled wave structure sets. The discovery of this material effect which was already known in a similar form in ancient Japan as *chirimenjifu* became Gisela Progin's trademark.

The strength of her clothes was in the simplicity of the cut. She really understood about letting material be material, making paper be felt as paper and by covering the body with scarcely noticeable, minimal cuts, preserving the originality of the material.

In addition to clothing, Gisela Progin also developed accessories like hats, bags or jewellery which she knitted from shifu yarn. In her last phase she also worked intensively with whole Nepalese sheets of paper which she crumpled like kamiko, creased, glued, sewed like material or draped to form silhouettes or design objects. She imported large quantities of the Nepalese paper as well as shifu materials and yarns and began to market them in Switzerland.

Gisela Progin became the most well-known shifu expert in Switzerland. Although she was not a weaver herself, she was able to bring its traits closer to others and through her infectious enthusiasm, and lectures and courses, she opened up this special world to many people interested in textiles.

Gisela Progin was an artist through and through. She was always looking for herself and 'her' themes, preferably travelling and always changing. In the middle of her creative bloom, she died tragically on a ski tour in April 2002. Her individual personality cannot be replaced, but her love for shifu has made waves. Many impulses which started with her, live on and are a monument to her, not least with this book in which a contribution from her should have been published.

Kamiko-like clothes

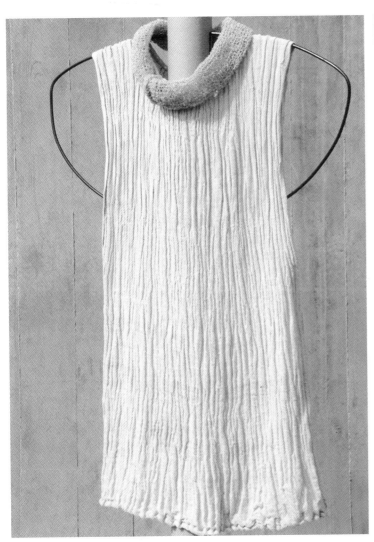

Mäti Müller

There is probably no one else today in the German-speaking world who knows so much about shifu and who pursues it with such seriousness and continuity as the weaver Mäti Müller. In her work, she tries to unify apparent contrasts in a playful way. As the only Swiss weaver, she has been working with industrially-produced paper yarns as well as the Japanese shifu tradition and has therefore forged a link between the two cultures.

Mäti Müller (1945) was a full-time youth worker for many years. When travelling and on seminars in Scandinavian counties, she discovered her love of the North and trained her eye for what she does today. Mäti Müller has always been fascinated by textiles. In addition to her job, she examined textile processes and design basics with increasing intensity. She took courses at the Design School in Zürich and finally set up her own workshop in 1976.

In 1979 Mäti Müller received a scholarship to the University of Textiles in Helsinki to study weaving from scratch. The time she spent studying reinforced her desire to concentrate on weaving as her main profression. She continued to return to the North for further training.

Mäti Müller has been living and working for several years now in a remote, mountain village in Calancatal. Here, a place which is only accessible by foot or with the funicular, she set up her 'papier-t-raum-matten' (paper dream meadows) workshop where she works on a freelance basis all year round apart from a few months in winter. This area fascinates Mäti Müller because of its complete seclusion, concentration on the essential, the traces of the needs of the elements and the cycles of nature.

Mäti Müller became acquainted with European paper yarn during her studies in Finland. The material made an immediate impact on her and she began to experiment while she was still in the North. She has been working consistently with the classic strings in her own very personal shape language for many years now. Her trademark became the so-called island peak pattern, a special and quite time-consuming form of twill binding. Two different twill movements are combined to produce diagonally-running, triangular or diamond-shaped patterns. The pretty, geometrically-reduced shapes have a clear patterning and join the surfaces in an interesting way. To make carpets, the weaver usually uses natural-brown paper strings in combination with a black cotton warp and black carpet tape. The colour selection is always bright and merry for table sets. In one series, the colours and shapes are structured in such a way that depending on how you combine them on the table, different patterns are produced. Carpets, table sets and sometimes wall objects made from Finnish paper strings are usually commissions. In addition to these paper weaves which are mainly for selling, Mäti Müller produces other things too: she was one of the first people in Switzerland to fall in love with shifu and is now regarded as an expert in this field. She first read about washi and the possibilities it offered for making paper yarn in 1988 and she was instantly fascinated. Paper as an independent medium for her own creation had always attracted her, but she wanted to remain faithful to weaving. She now saw the possibility of combining both of her interests. She slowly tracked it down, researched text books, set up contacts with Japanese people or people who were travelling in Japan and they sent her information and original material. She gradually acquired a clearer picture of a long, extremely polished and deep tradition. It occupied her more and more and finally she began her first own shifu attempts.

In 1993, the Japanese paper masters Yoichi and Mieko Fujimori from Shikoku (AWA factory) gave a papermaking course in Switzerland where they taught the original, Japanese method. Mäti Müller took part and came much closer to the secret of good shifu papers and the production of threads. She now makes some of the papers for her materials herself and almost understands the entire, time-consuming process of traditional shifu-making from her own papermaking experience.

When Mäti Müller heard about Gisela Progin's trip to Japan in 1993/94, to research the shifu tradition there, she followed her work closely. At a lecture, held later by Gisela Progin, the two artists met and an active exchange began. They organised conferences on the subject and shifu workshops and they invited the Nepalese weaver, Deepak Shresta to come to Switzerland to talk about his craft.

In addition to using her own and Japanese paper, Mäti Müller also began to experiment with Nepalese papers. She wove shifu threads into (usually) white, silk warps. This often produced double weaves. She then inserted Japanese or Nepalese paper between the two layers. These were cut to various geometrical or concrete shapes to fit a certain theme. This produced very delicate, airy room dividers, window or wall materials with a wonderful texture which live from the interplay between light and shadow and balance out the heavily coloured material made from industrial paper strings nicely. Mäti Müller wanted to achieve effects which are typical for this special material and leave it up to them in her woven interior furnishings textiles. In her most recent work, she has been trying to dye the paper with plant dye to test out further possibilities.

Mäti Müller now lectures both at home and abroad and gives courses on paper weaving and shifu. Her skill is well-known by experts in her field and a small shifu network has built up around her.

In Mäti Müller's work, it is clear the that much-quoted creative principle 'less is more' has a deeper resonance if the person behind the work has declared these words to be their whole life principle. As she understands it, knowledge about background and history are important prerequisites for one's own actions. According to her, only once you have understood how tools and material react, can you go beyond them to become creative and active. Only then can you follow a route which veers between learning, accepting, trying out and action. Mäti Müller proves impressively with her contemporary materials that she has mastered this balance, that she is a tightrope artist between East and West, tradition and innovation, and applied and artistic.

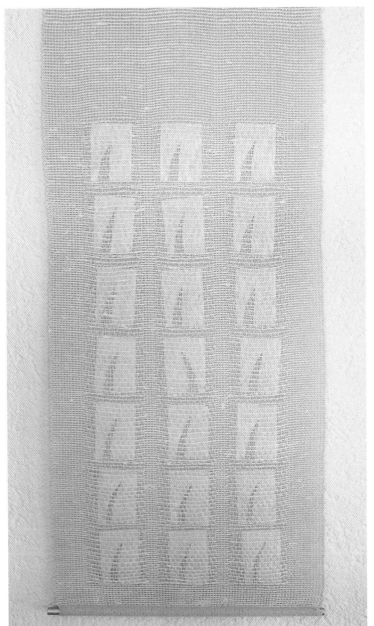

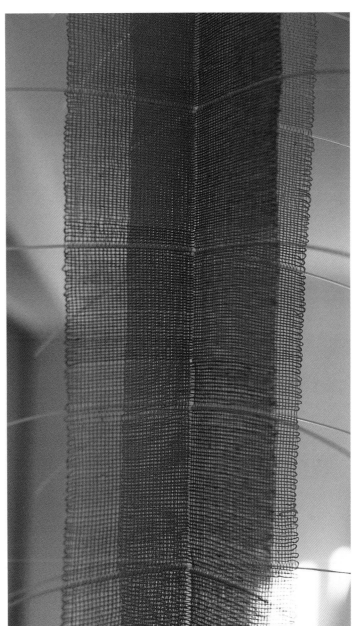

Lis Surbeck

Lis Surbeck (1952) is an extraordinary weaver whose fascination with shifu thread still has not wavered after several years. After her classical hand weaving training, she produced mainly consumer textiles. When she began weaving again after a long period of maternity leave, she approached the craft from a different angle. Her work was less determined by the end product, its function and the most rational way of making it, and more by how it came into being, its material aspect and the reduced weave of the threads.

Lis Surbeck discovered paper as a textile material in 1997 at a workshop led by Mäti Müller and Gisela Progin and she took a further education course on the same subject in Bern. Her enthusiasm for the costly process and the preciousness of the shifu material continued to grow. She began to examine it intensively on her own and soon found her own individual means of expression. 'I was fascinated by the transformation of a sheet of paper into a thread. These transformation steps are like a meditation or a metamorphosis of a plant or a butterfly. It is a wonder every time to hold a piece of material in your hands that you can wear and take care of. And you know that this piece of material carries the preciousness of the transformation process inside it.' Her deep respect for this process is palpable in all of her work.

Lis Surbeck usually spins her shifu threads from kozo or loktha paper, but she also experiments with other qualities, for example, papers from Bhutan or recycling material. She now gives courses too and has developed her own hand spindle for this so that she does not need to be sat at a spinning wheel when teaching. Lis Surbeck has been leading her own weaving workshop again since 1998. Here she produces very precisely worked, simple materials, scarves and so-called 'curtain pictures' for rooms. These are double weaves with opening side pockets into which different inserts can be pushed – leaves from nature, sheets of paper, threads or little lead tiles. These can be continually varied, producing an interesting interaction between foreground and background, transparency and density. In addition to her fascination for double weaving, she has been working primarily with all varieties of plain weave, which accommodates the extraordinary material language of the shifu thread particularly well. She never seems to be bored with her skilfully selected variations. For example, she plays with woven paper strips which have only been twisted into a thread in some areas to produce free, organic shapes out of dense and thin areas. Or she arranges the typical shifu bobbles in the weave deliberately according to a certain system, by fitting the width of the sheet of paper exactly to the width of the material, producing surprising, almost ikat-like patterns. These effects preserve the originality of the paper and can be achieved with almost no other material. The materials are almost always done in light natural tones. She sells her high-quality scarves and uniques pieces for the home at group and individual exhibitions or in Swiss galleries.

Lis Surbeck's shifu materials are clear and are not overpowering. The design follows the material and for this reason seems so coherent/ordered/at one with itself. You can feel the respect for tradition in every weft, but the weaves are not frozen in reproductions which are true to the original, they remain open to experimentation and the perpetual transformation which this material carries in it. Technical perfection and a love of fine things seem to flow automatically from Lis Surbeck. With great conviction, she circumnavigates the laws of productivity and its destructive effect – that is pure luxury, plain and simple.

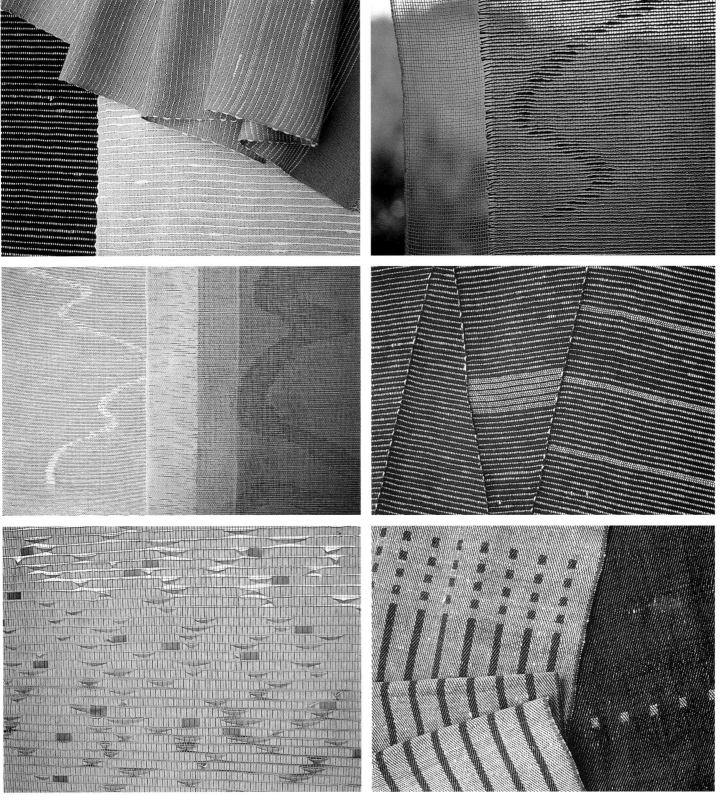

The Teppich-art Team

Craft and drawing teacher, Hugo Zumbühl (on the left in the picture) and the weaving master, Peter Birsfelder (on the right) met in 1998 at a course given by Mäti Müller on weaving with paper yarns. This was the beginning of an extremely successful team.

Hugo Zumbühl is a passionate designer. He spent six years in Peru where he worked as a development aid assistant and technical adviser in a weaving cooperative. The feeling for form and colour of the folk art there left a permanent impression on him, but he also counts the Japanese aesthetic amongst his big ideals. Hugo Zumbühl has always been impressed by paper as a material. When he returned to Switzerland, he began to experiment with different paper qualities in his small workshop in Felsberg and he worked them on his hand loom. He loves the reduction process and the rigid limits of plain weave, which directs the attention to the truly essential creative criteria. Boxes full of interesting material samples in his workshop provide proof of how many new and coherent variations are really possible from what seems such a trivial meeting between warp and weft. He uses these to try out different paper qualities, thread ratios and different surface treatments.

To be able to implement his ideas on a larger scale, Hugo Zumbühl needed a larger workshop and detailed mechanical and technical weaving knowledge. He therefore met Peter Birsfelder at just the right time. Peter had been running the weaving workshops in the Thorberg penal institution for 15 years and was looking for a meaningful, contemporary alternative to the classic rugs which the prisoners had been making there for years. The search for innovative design and and suitable possibilities of achieving it, connected the two passionate textile artists immediately and they set up the Teppich-art (carpet art) team within a few days. Hugo Zumbühl came up with the designs and was responsible for presentation and marketing, Peter Birsfelder developed, together with him, a series of products from the designs which could be made. They were then made using the heavy hand looms with the help of the prisoners.

In 1999 the Teppich-art-Team produced their first collection of rugs. It consisted of five compact models in rep weave. Thick paper cords or acid-free crepe paper from the packing industry were woven into a strong hemp warp which would become invisible later. This material can be cast on and pushed together particularly strongly so that the crepe paper stretches to enclose the warp thread and produces a sturdy, grainy surface. Because of this thickening process during weaving, the materials are very vivid and are about one centimeter thick. The edges of the rug are not trimmed with a border, they have a natural woven edge. At the beginning and end of the material, the warp threads which are sticking out after removal from the loom are simply pushed in and glued, so that four, pretty, simply outer edges are produced and the entire power of the weave is visible as a cross section. Both sides of all models can therefore be used.

On some carpets, the paper strips are dyed before weaving, but today, because of cost implications, colour is usually applied to the finished weave. This produces, for example, the deep, endless black effect of the 'Oscuro' model. The paper is worked wet and slowly dried over several weeks. Because the weave shrinks when it dries and the warp threads at the end have to be gone over again, you need to calculate an excess width when you are working, so that you obtain the required size at the end. A long phase of product development was needed before all of these criteria fitted in with one another.

The rugs of the Teppich-art-Team are impressive

because of their simplicity and almost archaic effect. The design as well as the material and the working process is long-lasting, has a promising future and exudes strength and elegance.The rugs can fill entire rooms, but can also take a back seat.

To obtain a particularly stylish shine and to cover the rug with a natural protective layer, most of the models are treated with natural wax at the end. This is rubbed into the surface with brushes or a spraying device and it is then polished. Rich, exceedingly thick and enclosed surfaces which are extremely pleasant to walk on are the end result. The effect is almost like that of a mixture of leather and silk. The products from the Teppich-art-Team are long-lasting, but they need to be taken care of and waxed every now and then. They are collectors' pieces, because despite the relative cheap working conditions in the prison, the products are relatively expensive. Because of the many steps which are done by hand, making the rugs is extremely work-intensive. A total of two working days are required to make one quadrat metre.

Since the beginning of 2003, various models by the team have been produced by the renowned Swiss carpet firm, Ruckstuhl, under the name 'Kollektion plus' (Collection plus). The innovative craft products of these two textile artists can now be bought from about 100 point of sale throughout Europe and the US.

The latest rug models from the team play with other experimental materials: with the rubber tubing from old bicycles, cut up pieces of woollen felt from old military blankets, or native goat hair. Another interior design object was also developed, the sophisticated 'transparavent', a mobile, light-pervious weave made from linen, paper string and paraffin-treated silk paper. With the help of woven cardboard strips, the material stands on separate metal supports so that this roomd divider can be flexibly varied according to each situation. The Teppich-art-Team have already been awarded several prizes for their meaningful and thoroughly aesthetic form of recycling. Amongst others, they have received the Swiss Design prize and have been awarded the the Swiss Shape Forum prize three times. Hugo Zumbühl and Peter Birsfelder guarantee exclusive, first-hand design with a forward-looking, ecological and socio-political background, but even if you do not know them, the models will convince you, because the rugs speak for themselves.

The 'transparavent'

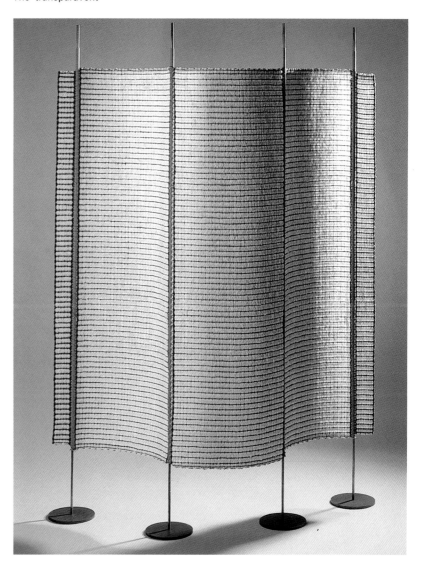

The jet-black rug, 'Oscuro'

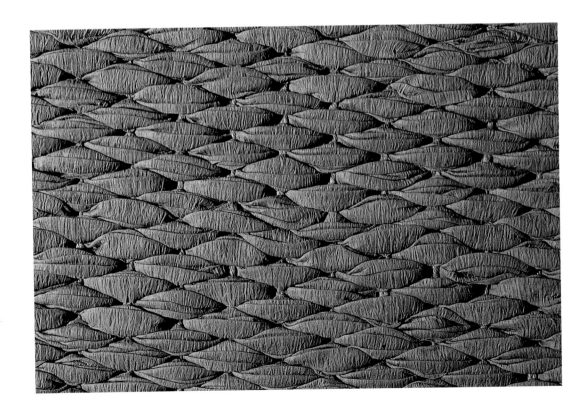

.5 Austria

Linda Thalmann

Linda Thalmann is currently studying at Linz University of Art (Textile and Design Institute). As a child, she discovered her love of paper making. She worked out various paper-making processes from reading about them and experimented with different native plant fibres and took relevant courses. She became acquainted with paper in the form of industrially-produced yarns for the first time during her studies.

Fascinated by these newly-discovered possibilities of the weaving technique, she began to experiment with different weaves on a 24-shaft hand loom and, in so doing, cleverly used the special characteristics of paper strings. Her first collection of samples are all natural white and they play with the combinations of different materials. For example, stiff paper yarns are 'hidden' in weaves which consist primarily of unspun, shining silk, flowing cotton threads or wool. In this way, Linda Thalmann achieves surprising material qualities which display completely different characteristics than those which at first glance would lead you to expect. Other samples play with the plasiticity of the strings. She thickens and loosens the rhythm of casting on to obtain stiffer and more fluid sections. Or she cuts floating threads up to let them stick up like bristles from the two-dimensional surface. 'I like the stiff, stubborn character of the paper string. It tells you exactly what you can do with it and what if refuses to do, there you cannot compromise with it! It seems stubborn and non-conformist – that's probably why I like it, because it's just like me!'

In addition to paper yarns she has bought, Linda Thalmann also works papers into her weaves which she has made herself. Strips from robust abaca, hemp or cellulose papers give new structures. Materials which are wettened, folded and pressed after weaving are particularly surprising because of their bark-like creased folds and the very plastic crash effects which nonetheless appear natural. It is the inexhaustible possibilities which the

material offers that Linda Thalmann finds so attractive. She wants to make exciting, material structures with a papery character. Areas of application and criteria for practical use are still, temporarily at least, less important.

In the summer of 2002 Linda Thalmann also learnt about the shifu process at a course given by Mäti Müller and she has since been combining both types of yarn in her work. On her many travels, she enjoys digging up unusual papers, to find out about their inner structure so that she can later use them in her textiles as cut strips or twisted threads where they can show where they have come from. In addition to material, she also makes three-dimensional art and book objects from handmade paper. She layers many sheets over or behind one another and even uses twisted strings from the base material to create effects. By inserting threads into the papery surface, she produces relief-like surface structures.

Linda Thalmann's latest pieces combine the fine European paper strings with felt. She lays loose pieces of thread, very loosely worked knitwork or chain stitches on the layers of wool and felts them together. The woollen fleece gets caught up in itself during this process and shrinks greatly, but the paper string does not. It maintains its full length and displays its very stubborn, robust side and begins to crimp. The unharnessed, randomly lying threads produce a dynamic, graphic structure and contrast well with the compact, fleecy background, part of which is then dyed with natural dyes.

These works make it very clear that Linda Thalmann's love is the fibres. When making felt, a compact surface is simply produced (in the same way as papermaking), by

joining the individual elements together directly. She treats the pieces of paper string that she has worked in like separate, stubborn fibres. She gives them unlimited space so their character can fill the space. In this way, she is very near to her original access point to the subject of paper, i.e. creating from the great variety of nature's elements.

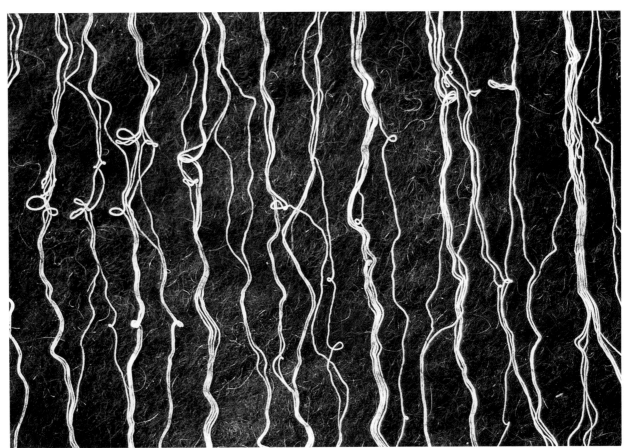

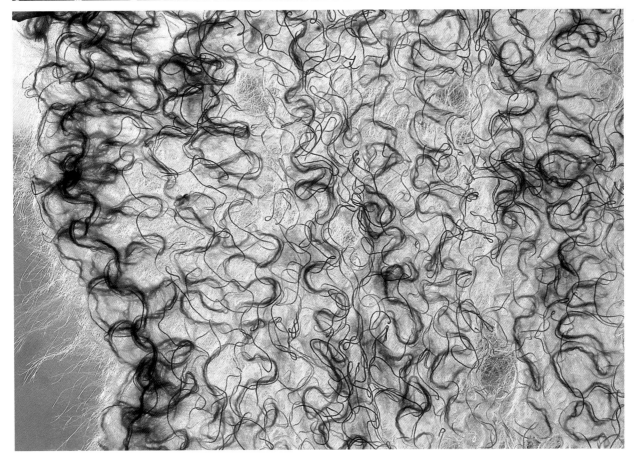

4.6   The Netherlands

Marian de Graaff

Marian de Graaff, from Holland, studied textiles at the Design Academy in Eindhoven from 1984 to 1990. In her work, she has specialised in researching unusual textile materials. She became acquainted with paper yarns whilst she was still studying. She has been fascinated by them ever since, testing their possibilities and limits using different weaving and entwining techniques and enjoying the surprises they still have to offer.

'I want to develop materials which are full of character. I most enjoy working with unusual materials like jute, coconut, steel, seagrass, sisal, rubber, and especially paper. These are troublesome, stiff, unconventional materials which are a challenge to work with.' Because they are so difficult to manage and require a perfect technique, she loves harnessing them with sophisticated working methods, yet still giving them their independent character space.

The designer achieved the meaningful structures in her numerous experiments primarily by combining different materials like wool and paper, metal and sisal, seegrass and paper, rubber and linen. This produces exciting contrasts as matt and shiny, rough and smooth, transparent and opaque areas clash within one textile area. For Marian de Graaff, the material is the message and the basis for every creative idea. Many of her designs are thought up while she is working, consciously listening to the naturalness and self-evident truth of the material. Her greatest aim is that her creations carry this natural, self-explanatory aspect within them.

Twenty years ago, Marian de Graaff also discovered her passion for macrophotography. She always has her camera with her when she takes her numerous hikes along the coast and on her travels in the rugged landscapes of Scandinavia. She collects different natural objects to analyse their formal laws at home. She is impressed by the fundamental, microscopic shapes in the environment, by the subtle shades of colour and the structures which have formed over years. The ordered caprices of nature are reflected in her textiles.

Marian de Graaff also designs for different industrial companies as well as being a freelance designer in her own workshop. In 2000, she received a working grant from the Dutch government's Fine Arts Fund to research the textile possibilities of paper. She was then able to devote herself to her experiments completely and she produced a whole series of exciting textiles from paper yarn, sometimes combined with linen or cotton. She presented these in an exhibition in Eindhoven in 2002. Most of the unique pieces were made by hand using different crochet and knitting techniques. Soft, fluid structures can be produced from relatively stiff paper strings using these entwining processes. It would be almost impossible to achieve the same effects from weaving. The textile character is further increased with follow-up treatment methods like washing, pressing or oiling. This makes the structures washable and pleasant to wear. In addition to stitched structures, she also presented weaves made from paper and cotton with interesting, wave-like surface structures.

The decorative interior furnishings textiles and the soft, coiling scarves were very well received by the public because of its novelty and technically perfect execution. The designer now hopes that companies will become involved with paper string experiments and its designs too. Marian de Graaff has already received a lot of prizes and grants for her special materials and has exhibited her work at numerous exhibitions at home and abroad. It is noticeable in her work that the technical possibilities from working with paper string are by no means exhausted and the material has a great future ahead of it, also with regard to the industry.

4.7 Denmark

Ann Schmidt-Christensen   Grethe Wittrock

Project Papermoon

Grethe Wittrock (on the right) and Ann Schmidt-Christensen (on the left) from Denmark founded an unusual artists' community in 1993 which made use of the versatility and power of expression of paper materials to make innovative, conceptional clothing. They called it Project Papermoon.

The two designers studied at the Danish School of Art & Design in Copenhagen between 1987 and 1993. Ann Schmidt-Christensen studied fashion and Grethe Wittrock studied textiles. Grethe Wittrock, who had already completed a classical apprenticeship in weaving before she began her studies, was given the opportunity to study in Japan for a year in the textile class at Seika University, College of Fine Art in Kyoto as part of a study exchange programme. She familiarised herself with the great Japanese paper tradition here and also became acquainted with the classical Shifu tradition. She was impressed by the great meaning of this tradition, but knew that she could not have used the precious material for her work because of cultural and economic reasons.

By chance, she found a Japanese company who produced industrially-manufactured paper yarns just before she was due to go home. These immediately inspired her desire to create. The materials were made from abaca and ramie fibres, two traditional, Asian papermaking plants. Because they are made industrially, these yarns cost a fraction of the handmade shifu material. but they still have an 'Asian' effect because of their unusual source fibres. Although the fibres do not display an irregular, burled structure, they are more supple and transparent than the compact, European paper strings made from wood fibres. Very different qualities of yarn can be obtained from these papers depending on thickness, width of strips and the power/thickness of twisting. Sometimes untwisted strips are used, sometimes the paper is decorated with a silk or ramie thread core to increase

stability. Some yarns are waterproofed, waxed or perforated with a substance enriched with viscose fibres, to produce compact to almost transparent materials.

Once she had packed large amounts of these yarns, Grethe Wittrock returned to Denmark and began to test the different qualities systematically in her workshop. She produced materials with very different densities and weaves on her hand loom and she went through all the varieties of surface finishings on the finished samples. This intensive work further reinforced her enthusiasm for the versatility of the material and she began to want to apply her materials in greater style. So she renewed her contact with Ann Schmidt-Christensen in 1993, whom she had known whilst studying. The fashion designer was immediately taken by the particular charm of the paper materials and saw them as the perfect material to use in her visionary ideas of fashion. Project Papermoon was born.

The Project Papermoon models were produced together by the two designers. Selected material samples made by Grethe Wittrock inspired extravagant designs in Ann Schmidt-Christensen. Together they established a conceptual meaning, got to the bottom of the technical details and discussed possible finishings.

All the materials continued to be produced by Grethe Wittrock on the hand loom in her workshop. The designer had been back to Japan several times since her study visit to choose material and order what she liked. Her repertoire now consisted of over 30 different qualities, some of which were specially made for her. She originally used 100% of the paper material for the warp and weft, but newer materials also play with unusual combinations of material. For example, fleecy mohair, wool or cotton tape are contrasted with simple, smooth paper. The weaves are technically perfect and before they are cut to size, they are washed so that they are softer and more absorbant.

Ann Schmidt-Christensen understands about giving her time and attention to the peculiarity of the material in her designs to lend it a new dimension. The patterns are very simple, often just consisting of strict geometrical shapes or individual elements which are transformed into three-dimensional body coverings by clever folding. The Japanese models are visible in her work.

What weaving has achieved for the desired pattern shape, Grethe Wittrock takes over for making the material. The weaving width is adapted individually to each pattern, where required, loops, slits or fringeing is worked in so that the material changes as little as possible and just needs to be 'assembled' more. So even before the material exists, the piece of clothing has already been thought out and the technical pattern has been exactly calculated. In addition to the weaves, Ann Schmidt-Christensen also knits paper yarn clothes on her knitting machine.

A large number of the pieces of clothing are made in white, further increasing the simple, almost futuristic character of the objects. Only the detailing is silver, black or grey. Different surface treatments like dyeing, printing, painting and cold-pressing are sometimes carried out by Grethe Wittrick on material which has not yet been worked and sometimes on finished articles. What she uses matches the established theme of each unique item. The two designers find their source of inspiration for these meanings by working with human habitats, foreign cultures and traditions and in nature's shapes.

Projekt Papermoon sees itself as an experimental research project in the clothing design sector, aesthetically, technically and functionally. They want the possibilities and limits of fashion to be discussed and guess at what fashion could be in the future. Clothing as communication over identity as the interface between our inner and outer world. In spite of this design concept, the fashion from Project Papermoon is extremely wearable. The shapes are technically perfect in their cuts or pleats and fit the body

The Kimono

perfectly. Even the material has pleasant qualities for wearing it – from stiff, stably-shaped weaves to flexible, fluid, almost transparent knits. This fashion is light, keeps the heat in, washable and makes a pleasant rustling sound when you move. From the very beginning, Project Papermoon's clothes met with great public interest. The artists were included in a cultural support programme led by the Danish government and were invited to guest lectures at art academies. Individual unique pieces or the entire collection were already represented at many exhibitions in galleries and museums beyond Denmark, for example, in Germany, France, Great Britain, Scandinavia, but also Japan and South America. Although Grethe Wittrock and Ann Schmidt-Christensen received many honours for their innovative work, they can barely live from Project Papermoon. The clothes are too experimental and because of the huge amount of manual work involved, they are probably also too expensive to be suitable for sale. The

Gallery

aim of both the designers is not to meet the tastes of the public, but to give shape to their common visions. Their joint fascination for the very special language of a material binds them in this aim.

Folded Dress

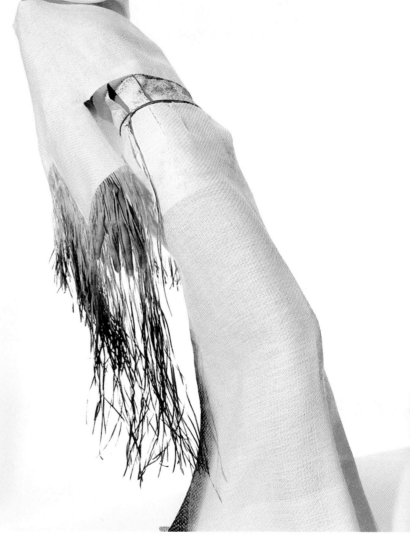

The Polar Bear

The Bird

4.8 Finland

Ritva Puotila and the Woodnotes company

Ritva Puotila, from Finland, (1935) is one of the big names on the international textile scene. She was the first to rediscover the wartime paper string as a material in its own right/with its own powers of expression and therefore introduced the come-back of the former substitute material. The innovative interior furnishings textiles and accessories made from paper string by her own company, Woodnotes, have been inspiring textile designers across the whole of Europe since 1980 and have triggered a real boom for this material.

Ritva Puotila is from a farming family and she grew up in the endless expanses of untouched, Finnish landscape. She studied set design, but never worked in this field because she became increasingly interested in the design of everyday things whilst she was completing her training and she tried out different techniques and materials. She soon began to design and weave ryas, traditional Finnish wall coverings which are knotted from brightly-coloured wool. At 25, as a young representative for Finland at the Milan Biennale, she received the gold medal. This was the start of her remarkable career.

At this time, in the 1960s and 1970s, Finnish design was flourishing. Many designers achieved international fame and within the country itself, design received considerable attention which inspired a whole generation of young art students. Ritva Puotila was able to profit from this and quickly made a name for herself as an artist and designer. She soon became known in the field for her inimitable feel for subtle colour combinations and her simple 'Nordic' aesthetic. She named the simple Finnish farming culture and the uniqueness of Scandinavian nature as her source of inspiration. Numerous trips to East-Asian, African and South American countries also influenced her view of things over the course of time.

As a young wife and mother, Ritva Puotila made a conscious decision to work freelance so that she could divide up her time herself and so that she could work from home. In the 'Ritva Puotila studio', an old house near Helsinki, in the middle of nature, she consistently produced unique textile pieces, but also industrial designs for well-known companies throughout Europe and the US. In addition to textiles for interior furnishings, she also produced crockery designs made from glass, ceramic and plastic.

One of the companies for whom Ritva Puotila designed, was Tampella, the biggest Finnish linen spinning and weaving factory. It had also made paper yarns for the Finnish textile industry during the war and in the following years. At the end of the 1970s, Ritva Puotila came into contact with this company with old paper string rests which were being used for isolation. She already knew the material from her childhood, but she had forgotten all about it. She now saw the strings through different eyes. Impressed by their special, harsh flair, she established that their special qualities like stiffness, lightness, dust-resistance, and their simple, practical appearance would fit perfectly into modern times. She made her first experiments using knotting and weaving techniques with the available left over paper strings from Tampella.

The initial results of her rug and table weaves were promising and showed something completely new and unmistakable. It also became clear, however, that with regard to the material quality, a lot of development work was still required. The special power of expression of paper textiles had entranced the designer. She believed in the potential of the paper string and, after lengthy consideration and experimentation, she decided to invest in her idea. In 1987 her dream became reality and she founded the Woodnotes company with her youngest son, Mikka Puotila. It is an innovative company and it soon made a name for itself outside Finland too. Ritva Puotila's designs were made in a small factory in West Finland. They were the first series of rugs, table sets and room dividers to be woven from paper yarn. They sold well straight away. In the first few years, she worked intensively trying to improve the yarn quality and on ecological dyeing methods. Tampella had started to make large quantities of yarn again and the quality of the war material was soon far exceeded in terms of being tear-proof, its surface structure and the non-fade properties of the colours. When Tampella had to close in 1992, Woodnotes bought the old paper spinning machines and began to produce the yarn themselves. Because demand was constantly growing, a year later the Suomen Paperilanka Oy company took the production over and is still producing in a big way today. This then gave other firms and textile designers access to the material.

The first products made by Woodnotes (which are also the most sought-after today) were paper carpets. The yarn is woven into a strong cotton warp in rep weave. The strict, geometrical pattern is always done in high-quality natural tones, although a rich colour is always combined with natural brown. Proportion, colour and working methods fit together harmoniously and underline the aesthetic expressiveness of the material. All the other product series which were developed in the years that followed also continue this clear line. The team developed table sets, blinds, room dividers, cushion covers and different variations of bags from Ritva Puotila's designs. Some of the bags were made from 100% paper, some were made with a combination of linen or cotton. Upholstery materials were also made and a furniture designer was hired and she now designs simple chairs and sofas for Woodnotes. The products all have one thing in common: they radiate something special and stylish, a beauty which the material carries within it and which Ritva Puotila has awoken, as hardly anyone else can. 'We were the first in the world to make functional textiles from paper strings and not to imitate another material, but because of the beauty of the paper string itself.'

In addition to aesthetic criteria, social and ecological factors are also very important to the company. Their close link to nature is reflected in the company name. On the one hand, Ritva Puotila sees nature as a starting point for her paper and as a continual source of inspiration for creation. On the other hand, she sees it as something to be protected and which has to be taken into account in spite of all economic considerations. So all the wood for their paper string production comes from Finnish woods, all the steps of the production process are carried out by small, nearby, Finnish companies and dyeing and bleaching processes meet the strictest ecological standards. The product quality is also considerably increased by the fact that the textiles can be burnt or composted without releasing harmful substances. They are dirt-resistant, dust-resistant and are therefore ideal for those with allergies.

Today, Woodnotes is a large company with 60 employees and they sell over 90% of what they make in 50 countries across the world. At the beginning, the company was the only one selling paper string articles. Since then, other paper spinning factories have started up again and

competition has increased in the last few years. More and more rug makers have been using the rediscovered material themselves.

So that they can still maintain their hold on the market and preserve their own touch, in addition to industrially-manufactured products, Woodnotes has also been making unique pieces and small series for demanding customers. The 'Ritva Puotila studio' where handmade design objects are made from paper string, was integrated into Woodnotes and they now sell exclusive wall designs and limited editions which cannot be imitated by other companies. Ritva Puotila designs materials which can only be made using a hand loom because they have complicated weaves or material combinations or are made using the traditional rya technique. With this knotwork, where the length of the pile can be up to 25cm (10in.), Ritva Puotila refers to the long, Finnish textile tradition and to her own creative beginnings as an artist.

The most recent work by Ritva Puotila exudes great lightness. The colours are lighter and clear. Radiant lemon yellow, fresh aqua blue, brilliant pink or grass green are the antithesis of the muted, rich tones of her mass-produced collections. The earth element has been expanded by adding the air element. The colourful yarns are very often worked into a metal warp which makes completely new effects possible. On the one hand, by combining pure, almost transparent metal weaves with thick coloured strips of paper yarn a nice contrast can be achieved. On the other hand, the metal warp allows the finished weave to be bent into exciting three-dimensional objects. For example, strips worked in double weave can be cut up like slats or bent to produce a choppy, plastic, surface structure. Or structural design objects are produced which almost seem to float in spite of their heavy weight.

Ritva Puotila has succeeded in being taken seriously as both an artist and a designer. Her pieces of art are exhibited in museums throughout the world, for example in the Helsinki Design Museum, the Victoria & Albert Museum in London, and the Museum of Modern Art in New York. Important public buildings like the foyer of the European Council in Brussel have been furnished with pieces of her art. The link between application and art in her creations has been acknowledged with numerous

prizes, honours and titles. This was also the key to the success of Woodnotes. According to Ritva Puotila, art also comes from ability, from disciplined work on a particular idea. She has been proving that she has this ability impressively and consistently for several decades. Like virtually no other, she can claim the paper string as hers.

4.9 Industrial manufacture

In the last few years, paper yarns are being used increasingly in the textile industry sector, although the material is only worked in the weft in machine weaving processes. Many rug companies have followed in Woodnotes' footsteps and started to make rep weaves from the strings, with very varied results. Sometimes, the Finnish models are simply copies, sometimes new, interesting qualities are produced. These almost always play with unusual combinations of material.

For example, in the Finnish company *Hanna Korvela*

Design Oy paper string, wool or popana (a traditional Finnish textile material made from strips of cotton knit) are woven alternately into a strong cotton warp to give a graphic pattern. In Switzerland, the traditional carpet company, Ruckstuhl, was one of the first to rediscover paper strings for use in their products and thereby found their own form of expression.They contrast the simple, practical material with dark, raw wool and light, shining viscose for example, or they combine sisal and paper from the same dyeing process to play with the subtle nuances of the different reflections of light. The paper yarns they use are from Scandinavia and are woven dry. The back of the weave is usually lined with rubber so that they do not slip and can be used to cover large areas of floor.

Hanna Korvela Design Oy (Finland): the Duetto and Encore rugs (Paper string, BW-Popana)

Weave with different densities of paper inserts, 100% paper, Sirpa Lutz (Switzerland); paper string object made from a carpet industry waste product

In addition to rug production, there are also some companies who make light textiles for interior furnishings and clothing materials and, in their continual search for new, innovative materials, they have begun to experiment with paper in the last few years. For the most part, they obtain the yarns from source which are inaccessible to private individuals. The yarns often come from the East-Asian low-wage countries. Every company keeps information about how they were made and exactly what materials were used to itself like a precious secret. Paper's good image is used in marketing and they play with the element of surprise that this material always triggers when it is used to make textiles, because people never guess that paper was used, but it corresponds to current trends.

For example, the Jakob Schlaepfer company in St. Gallen in Switzerland has been producing glamourous materials for renowned couturiers across the world for decades. They have recently started to use paper. Head designer, Martin Leuthold, developed a collection of materials in 1996. They are almost all made from recycled cardboard. The waste product is ground, it is sprayed onto special machines by jets where it is squeezed and twisted into fine threads. This new material is produced in Holland and woven in Swiss workshops. Similarly to papier maché, the paste makes it relatively hard and as tear-proof as strongly woven cotton. Sometimes the yarn is mixed with the usual natural fibres or with metal, plastic and lurex threads to produce unusual effects. This therefore produces new materials in extravagant weaves with a shining or transparent character, moiré effect or a strong structure. One of these weaves is particularly impressive because it displays a bark-like, wrinkled surface which is almost reminiscent of corrugated cardboard. Similarly to the Nepalese shifu materials, it consists of thickly woven, slightly overspun paper yarn in weft and comes completely flat off the loom. It only gathers and begins to form waves when it is wet. The upward folds are stiff and difficult to work with, but can be easily ironed out when it is moistened. This material which is bursting with character is particularly well suited to stylish, robust outerwear. When you look at it, it really is very hard to believe that for the most part, it has been made from rubbish!

The German material designer, Ulf Moritz also uses

Jacket by Esther Chabloz (Switzerland) made from wastepaper material from the Jakob Schlaepfer company (Switzerland)

paper for his latest creations. The much celebrated star of the textile scene live and works in Amsterdam and is well known for his innovative designs, subtle use of colour and his feel for unconventional, extravagant materials. He designs incredibly light and transparent high-tech materials which are created fro modern, light-filled rooms. As a freelance designer, he works for several companies and also has plans for furniture, wallpaper and glass products. He has been artistic head of the Nürnberg textile company, Sahco Hesslein, who market their own collections of textiles, since 1996. Finely cut small strips made from thin, but very compact papers are woven in combination with acetate, polyester, polyamide, metal, mohair or other exotic materials into fine, precious materials which almost appear to float. The synthetic fibres in the material give it a gleaming shine and allow the effect of the paper strips to be shown to advantage in the almost transparent weave

One of the most exciting developments in working with paper is probably what is happening in Japan. There is a company there which makes materials. This company

▶ The 'Karo' (checked) material by Ulf Moritz consists of 74% paper and 26% polyester.
▼ Ulf Moritz with Sahco Hesslein (Germany): various samples of material with paper in.

examines Japan's long textile and paper tradition in such an incredibly original and highly aesthetic way that it puts much of what Europe is currently up to in this sector in the shade. The Nuno company was founded in 1984 by textile designer, Raiko Sudo, who is still the creative focal point of the company. She develops materials which surprise because of their special, new material qualities. In a creative and completely natural way, she combines rich in tradition natural materials with synthetic fibres, or ones which are not used in textiles usually, such as metal, rubber or silicon. Nuno checks each material and each modern or traditional method for its aesthetic potential so that they can be used, if good enough. Because over 60% of the production is carried out in individual orders by small or family Japanese businesses, they are able to react relatively spontaneously to the newest ideas and implement technological developments immediately. The concept of the company is to exhaust the variety of expressive possibilities of textiles. Even the name Nuno, which only translates as 'material', indicates that the company produce textiles which are self-sufficient. The company keeps growing and there is now a branch in

Nuno (Japan): 'Hoshigaki' material

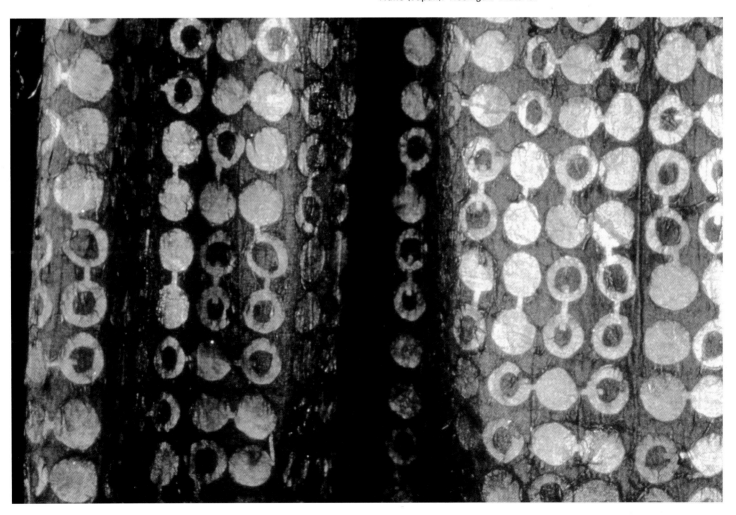

Switzerland which sells these Japanese treasures in Europe. In a wonderful volume of books, the Nuno company pay tribute to their creations with precious photos and the individual textiles are described in brief.

Amongst other materials, Nuno also use washi (traditional, handmade Japanese paper) for their creations. The region it comes from, which specific qualities it has and what it was used for traditionally – all of these are always very important in the creative process. For example, stable, kozo paper strips are woven between two transparent layers of weave made from silk organza. They are inserted to form a base material of flat, woven, floating paper strips producing a random, windswept surface, or circular and floral shapes made from washi are pressed onto a velvet underlay with synthetic glue. Nuno understand how to let the material characteristics, its stiffness, lightness and transparent density in the finished products express themselves best. The long and serious examination of the various qualities of the material can clearly be seen.

Other Japanese firms have also recently begun to use paper more in their repertoires. Even the great Japanese designer. Issey Miyake, has paid tribute to the great paper tradition of his country. As early as 1984, he created a whole collection of clothes made from strong, handmade washi, that was crumpled like kamiko and was treated with various oils to create draped wrap dresses and raincoats – clothes which are born of their historic roots but have visions for the future.

▲ Nuno (Japan): 'Slipstream' material
▼ Nuno (Japan): 'Patched Paper' material

176 | 177

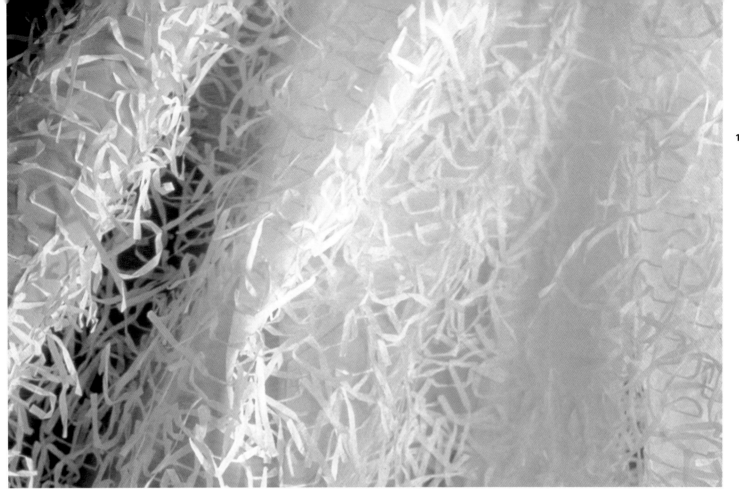

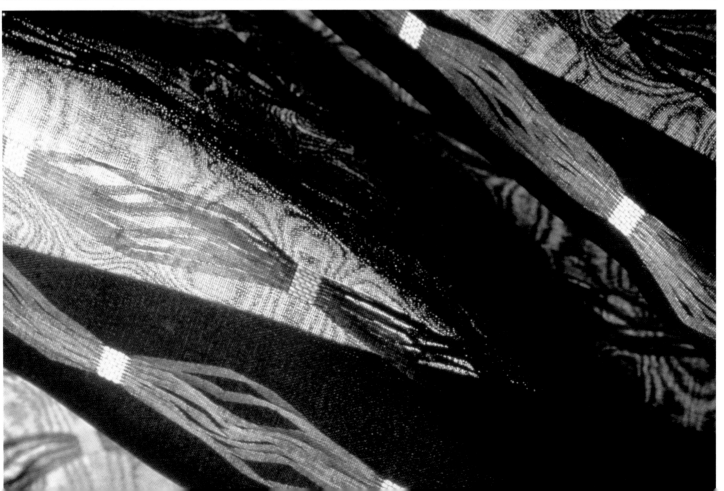

5   **Appendix**

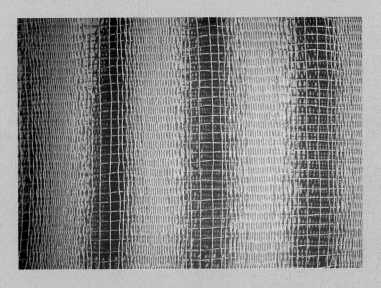

▲ Weave with different densities of paper inserts, 100% paper
◀ Shirpa Lutz (Switzerland) paper string object made from carpet
industry waste product

5.1   Timeline of Japanese history

Age/phase		Cultural and historical information	The history of paper
Primeval times	Jomon age (7500BC.–300 BC)		
	Yayoi age (300 BC–AD300)		
Age of court aristocracy	Yamato age (300–710)	From about 300: Shintoism, Japanese nature religion. The Yamato family clan ascends to imperialist race. 538: Buddhism from the Kingdom of Korea to Japan.	610: Chinese method of making paper is brought to Japan by Korean doctor and priest, Tam-Chi, Prince Regent Shotoku supports its development and arranges for mulberry busnes to planted across Japan.
	Nara age (710–794)	Founding of the capital, Nara, art and culture flourished. Building of central administration machinery following Chinese model.	Crucial improvements on Chinese methods. Paper is made by farmers in different provinces in Japan as a Winter occupation.
	Heian age (794–1185)	Transfer of capital to Heian-Kyo (Kyoto); development of Japanese script and literature; imperialist aristocracy in contrast to enormous poverty of the people.	Washi becomes a symbol for civilisation, purity and religious meaning, Kamiko materials first used in 11th century.
Age of warrior nobility (Shoguns' age)	Kamakura age (1185–1336)	Transfer of capital to Kamakura; economic growth, population growth, start of currency system, military feudalism led by Shoguns. Long period of domestic military conflict.	Washi becomes an irreplaceable material for wider population in very different areas of their daily lives.
	Muromachi age (1336–1573)	Domestic power struggles divide Japan into a Northern and Southern dynasty. 1549: Christianity comes to Japan.	Kamiko slowly gains popularity with higher.
	Momoyama age (1573–1615)	Government increasingly decentralised; European influences mark Japanese art.	Founding of first papermaking guilds.

	Edo age (=Tokugawa age) (1615–1868)	Transfer of capital to Edo (=Tokyo).Peace at last thanks to Shogun Tokugawa. 1639: Japan cuts itself off from abroad, Japan because the richest economic power in the Far East, extremely high technical level reached by much of population in designing and making craft products. 1853: Commodore Perry from America landed in Bay of Edo and forced opening up of the country, last Shogun abdicates.	Heyday of papermaking in Japan (paper as important source of tax income for government). 1638: First shifu materials made.
Modern times	Meiji age (1868–1912)	1868: Meiji reform: opening up of the country, constitution following Western model; start of industrialisation; cultural exchange with West. . 1894/95: Chinese-Japanese war 1904/05: Japanese-Russian war	Paper machines imported to Japan. Shifu and kamiko disappear from day-to-day living; in the course of Asia boom, Japanese paper becomes popular collector's item in West.
	Tasho age (1912–1925)	1914–1918: Japan is Germany's enemy in the First World War.	Industrial paper manufacture continues to drive out manual papermakers.
	Showa age (1925–1989)	1937: War against China 1941: Attack on Pearl Harbor; war against US 1945: Nuclear bombs dropped on Hirosihma and Nagasaki – Japanese capitulation 1947: New constitution, democratisation of Japan.	Washi is transformed from material for the masses to stylish article. Shifu and kamiko declared cultural heritage and production of them restarted, production of washi is supported by state.
	Heisei age (1989–?)	Traditional Japanese values mix with Western style of living.	Washi only still made by a few farmers for daily use; a few famous master papermakers and workshops gain international reputation and bring technique to West.

5.2 Glossary

a) Japanese terms

asa: collective term for bast fibres, obtained from bark, stems and leaves of Japanese plants (for example: banana fibres, hemp, ramie, Chinese nettle, etc)

asajifu: Shifu fabric with paper as weft material and a bast fibre warp.

chabaori: tea ceremony jacket; often made from Kamiko in the Edo age.

chirimenjifu: crepe-like Shifu weave, produced with the help of a slightly overspin thread.

fu: cloth, weave

fukucho: old accounts books, used by poorer members of the population to make shifu.

fusuma: partitions or sliding doors made from paper, applied to a wooden lattice.

gampi: plants whose fibres are used to make very fine and extravagant papers.

haiku: form of Japanese poem

haori: traditional Japanese, mid-calf-length men's coat

hirahaku: gold or silver plated paper lamella

hira-ori: woven in plain weave

jibaori: floorlength cape, often worn when travelling, usually made from Kamiko in the Edo age

kakishibu: Persimmon sap; tannic acid, obtained from the ebony plant and used for waterproofing when making kamiko (see also *shibu*)

kami: expression for a concrete sheet of paper; often written gami also; also homonym for the gods of nature in the Shintoist belief

kamiko: creased and waterproofed paper which is made into clothes

kami-shimo: shoulder dress and part of the ceremonial garment of the Samurai, often made from Shifu in the Edo age.

kesa: choir shirt of a Buddhist monk

keta: wooden frame with lid; part of the Japanese papermaking sieve

kinujifu: shifu weave with a silk warp and a paper weft

kobai-ori: rib-like shifu weave produced by changing the material in the weft

komon: typical Japanese pattern consisting of tiny, geometrical elements and symbolising certain natural phenomena

konnyaku: strengthening plant sap, extracted from the Amorphophallus konjac, tuber plant from the family of arum plants and used to waterproff Kamiko materials

koromo: monk's garment

koyori: shifu-like structure made from short, twisted paper strips

kozo: paper mulberry bush (*Broussonetia papyrifera*); most important plant in Japanese papermaking, often called kozu

memonjifu: another name for monjifu (mixed weave from cotton and paper)

menjifu: Shifu weave mix with cotton warp and paper weft

mitsumata: straw-like plant, native to Japan (*Edgeworthia papyrifera*), very elastic and shining papers can be made from its fibres

momigami: paper which has been made soft and elastic by crushing it; source material for making Kamiko

mon-ori: material with complicated weaves

morojifu: weaves where the warp and the weft are paper threads

nagashi-zuki: Japanese method of making paper where several layers are made on top of one another on a sieve

neri: thick, slimy substance, which is added to the fibre pulp to be able to make several layers and to prevent fibres and finished sheets from sticking together; obtained from the tororo-aoi root

obi: sash-like article of clothing. It is wound round and tied around the kimono.

Samurai: sword-carrying member of nobility, belonged to the knight class; carried out warrior tasks and seen by others as spiritual and moral leader.

shi: Japanese expression for paper in concrete or general terms.

shibu: persimmon sap, used for waterproofing when making kamiko and gives the paper a rust-brown to ochre-yellow colour, also often called kakishibu

shifu: weave, in which at least one thread system consists of spun threads made from Japanese paper.

shifugami: kozo paper, in which all fibres are brought to lie parallel using a special method; base material for making shifu.

Shintoism: old Japanese natural religion which believes that all things are living.

Shiroishi-Shifu: paper weave from the region surrounding the town of Shiroishi, known for its amazing quality.

Shogun: originally military title of office for great commanders; soon came to describe high members of the nobility, descended from the imperial family and held political power in the individual prefectures. Comparable to European princes.

shoji: traditional Japanese window made from handmade paper.

su: removable, rolling sieve made from tied together bamboo shoots.

suketa: Japanese papermaking sieve, consisting of a frame with a lid and removable, rolling bamboo sieve.

Sutra: holy script of Buddha

torafu: another name for kozo

tororo-aoi: type of hibiscus, neri is extracted from its roots. neri is a slimy substance which improves the consistency of pulp.

toyugami: another name for momigami

washi: collective name for all handmade Japanese papers (wa: Japanese, shi: paper) and all the spiritual background connected with it.

yorihaku: twisted flattened wires made from paper strips layered with precious metals.

yuu: fibres from the paper mulberry bush, which in addition to making paper, were formerly spun and woven like other bast fibres.

zaogami: high-quality paper from the Sendai region, usually from the town of Shiroishi

b) English technical terms

Amat: name for the tapa material of the Mayas and the Aztecs, also written Amatel or Amate

Bonded fibre fabric: non-woven fabric, consisting of at least 35% textile fibres

Caustic soda method: production of cellulose by boiling wood in caustic soda

Cellulon yarn: cellulose yarn where the strips are formed on a fixed mounted

Cellulose material: chemically opened fibrous mass, produced by boiling the wood in caustic soda, or a sulphate or sulphite leach

Cellulose: main component of the cell walls of leaves, stems, bark, petals and similar of almost all plants and base substance of wood, paper and many textile fibres

Cellulose yarn: paper yarn which is made when the fibrous mass is divided into strips during the papermaking process (also called 'paper material yarn')

Couching: act of putting the ready-made, but still wet, sheets of paper onto a pile of sheets which are each separated with a woolen felt cloth

Fibrils: macro-molecular particles, which make up fibres

Hollander: machine invented in about 1670 which chops up base substances, like plants or rags, for making paper and makes them into a pulp.

Hydratation: chemical reaction which triggers the withdrawal of water

Island peak: triangular pattern produced in a weave by the clash of two grades of twill going in different directions

Japanese gold: material used for effect, consisting of paper strips covered with precious metals (e.g. sheet gold) and mainly woven into brocade

Licella yarn: early name for paper yarn

Linen reform: term for items of linen (shirt collars, cuffs, undershirts etc.) made from waterproofed industrial paper and made current in the first half of the 19th century in Europe and America

Loktha: Nepalese daphne plant, used to make paper

Metric number/count: description of yarn strength, abbreviation: Mn; it expresses how many kilometres one kilogramme of yarn can run and is calculated by dividing the length of the yarn in metres by its weight in grams. The higher the number, the finer the yarn

Non-woven fabrics: synthetically produced, unwoven materials which are usually made from synthetic fibres, entangled and by glueing or melting made into a compact surface

Pextil Jersey: knit made from paper string

Polishing: polishing method where the surface of yarns is made smooth and shiny by adding wax, paraffin or similar

Pulp: viscous fibrous mass from which the sheet of paper is made, (also called 'suspension')

Rubbing: Producing a yarn by circular coiling and twisting of the flat paper strip

Rya rug: traditional Finnish knotted rug made from wool.

Shirting: thin, inferior material coating on paper linen

Serviteur: undershirt for men which only covers the front part of the chest

Silvalin yarn: cellulose yarn where the cellulose mase is separated into strips on the sieve by jets of water

Spinning web: Woven material which consists entirely of multi-filamented (endless) yarns, shaped into a surface thermoplastically

Sulphate method: production of cellulose by boiling wood in a sulphate leach

Sulphite process: production of cellulose by boiling wood in a sulphite leach

Surrogate: replacement material

Suspension: viscous fibrous pulp from which the sheets of paper are made

Tapa: bark raffia, produced from plant substances by storing them in stacks and hammering them. It has been used to make clothing, living room textiles or writing materials for thousands of years in different parts of the world

Textilite yarn: blended paper yarn, where paper strips are added separately to the textile fibres whilst spinning, or added to the threads which have already been spun.

Textilose: flat paper yarn covered with textile fibres

Textilose yarn: circular yarn produced from paper strips covered with textile fibres

Tyvek material: spun fleece made from 100% polyethylene, has similar optical characteristics to paper and is mainly used for costumes

Tub: large, usually wooden container, in which the pulp is made

Wet-shaping: method for making a non-woven fleece, produced hydrodynamically on a specially-developed type of paper machine

Wood grinding: fibrous mass is broken down mechanically, produced by grinding the wood

Xylolin yarn: early name for paper yarn

5.3 Classification of textile techniques

I Techniques for making threads

1) Producing threads — Drilling; Spinning; Twisting

2) Reinforcing threads — Twisting; Jasping

II Techniques for making material

1) Felting

2) Primary techniques for making material (without fixed warp system)

A) Stitch formation (with continuous thread)
   a) Hanging
   b) Entwining
   c) Knotting
   d) Lace stitching (Stitch, needle lacing)
   e) Crochet
   f) Knitting

B) Making material using thread systems (at least two thread groups)
   a) Partial braiding (one passive thread system)
      inserting; winding; binding, twist binding;
      Puffed partial braiding
   b) Real braiding (all thread groups are active)
      braiding parallel to edge; diagonal braiding; Zopf-,
      cord loop braiding; twist braiding; twist dividing
   c) Tatting
   d) Macramé

3) More advanced techniques for making material

C) Warp material method (with fixed warp system)
   a) Sprung (warp is active)
   b) Making material with a passive warp

D) Partial weaving (only one thread system can be raised)
   a) alternative partial weaving
   b) reserving partial weaving

E) Real weaving (at least two warp systems are active alternatively)
   a) Finger weaving
   b) Board weaving
   c) Grid weaving
   d) Eye bar weaving

III Techniques for decorating material

1) Decorating with additional elements whilst the material is being made
   A) Pile
   B) Beaded materials

2) Decorating the material after it has been made
   A) Appliqué
   B) Embroidery
   C) Decorating the material with liquid substances

.4 Bibliography

a) History of paper (general)

Bayerl, Günter; Pichol, Karl: *Papier. Produkt aus Lumpen, Holz und Wasser.* Reinbek bei Hamburg: Rowohlt, 1986

Berger, Dorit: *Die Geschichte des Papiers. In: Deutsches Textilforum.* Hannover: Scherrer. 4(1985), S. 16–17

Dardel, Kathrin: *Kreatives Papierschöpfen. Pflanzenpapiere. Recycling-papiere. Farbige Papiere.* Bern; Stuttgart; Wien: Paul Haupt, 1994

Dawson, Sophie: *Kunstwerkstatt Papier. Schöne Papiere schöpfen und gestalten.* Freiburg/Breisgau: Christophorus-Verlag, 1994

Franzke, Jürgen; von Stromer, Wolfgang (Hg.): *Zauberstoff Papier. Sechs Jahrhunderte Papier in Deutschland.* München: Hugendubel, 1990

Gale, Elisabeth: *From Fibres to Fabrics.* London: Allmann & Son, 1968

Hunter, Dard: *Papermaking. The history and technique of an ancient craft.* New York: Dover Publications, 1974

Katz, Casimir (Hg.): *Fachwörterbuch Papier.* Gernsbach: Deutscher Betriebswirte-Verlag, 1994

Pieske, Christa: *Das ABC des Luxuspapiers.* Berlin: Dietrich Reimer, 1983

Sandermann, Wilhelm: *Papier. Eine Kulturgeschichte.* (3. Aufl.). Berlin; Heidelberg: Springer-Verlag, 1997

Stümpel, Rolf (Hg.): *Papier.* Berlin: Museum für Verkehr und Technik, 1987

Thackeray, Beata: *Papier. Handschöpfen, Gestalten, Objekte und Skulp-turen.* München: Christian, 1998

The Scottish Society for Conservation & Restauration (Hg.): *Paper and Textiles. The common ground.* Preprints of the conference held at The Burrell Collection Glasgow. 19–20 September 1991. Edinburgh: Dupliquick, 1991

Turner, Silvie: *The book of fine paper.* London: Thames and Hudson, 1998

Trobas, Karl: *abc des Papiers. Die Kunst, Papier zu machen.* Graz: Druck- u. Verlagsanstalt, 1982

Webb, Sheila: *Paper. The continuous thread.* Ohio: The Cleveland Museum of Art, 1982

Weber, Therese: *Die Sprache des Papiers. Eine 2000-jährige Geschichte.* Bern; Stuttgart; Wien: Haupt-Verlag 2004

b) The Japanese paper tradition, Kamiko and Shifu

Adams, Edward: *Korean Folk & Art Craft.* Yongdong: Seoul International Publishing House, 1987

All Japan Handmade Washi Association (Hg.): *Handbook on the art of washi.* Tokyo: Wagami-do K.K., 1991

Arn, Ursina: *Shifu. In: Handwerk.* Altdorf: Gisler Druck AG, 3(1998), S. 19–22

Avitabile, Gunbild (Hg.): *Die Kunst des alten Japans.* Stuttgart: Edition Cantz, 1990

Barrett, Timothy: *Japanese Papermaking.* New York; Tokyo: John Weatherhill, 1983

Buisson, Dominique: *Japanische Papierkunst.* Paris: Terrail, 1991

Byrd, Susan: *Shifu. A Unique Cloth From Japan.* In: Ornament 12/2 (1988), S. 66–71

Byrd, Susan: *Shifu. Fine Handmade Paper Cloth.* In: Hand Papermaking. Vol. 1, Nr. 2 (1986), S. 18–22

Haks, Leo (Hg.): *Art Celestial. Paper offerings and textiles from China.* Ghent: Shoeck-Ducaju & Zoon, 1997

Hughes, Sukey: *Paper Clothing: A Brief History and Appreciation.* In: Chanoyu Quarterly. Tea and Arts of Japan. Santon, Charles (Hg.). 30(1982), S. 41–51

Hughes, Sukey: *Washi. The world of Japanese paper.* Tokyo; New York; San Francisco: Kodansha International, 1978

Jackson, Anna: *Japanese country textiles.* London: V & A Publications, 1997

Katakura, Norumitsu: *Japanese Paper and Kamiko, Shifu Fabric.* Shiroishi: Sendai Print, 1941

Kume, Yasuo: *Tesuki Washi Shuho. Fine handmade papers of Japan. Vol. 1.* Tokyo: Yushedo, 1980

Lübke, Anton: *Weltmacht Textil. Eine Wirtschaftsbiographie des Kleides.* Stuttgart: Veria Verlag, 1953

Maruyama, Nobuhiko: *Clothes of Samurai Warriors.* Kyoto: Kyoto Shoin, 1994

Minnich, Helen Benton: *Japanese Costume and makers of its elegant tradition.* Rutland; Tokyo: Charles E. Tuttler Co., 1963

Muraoka, Kageo; Okamura, Kichiemon: *Folk Arts and Crafts of Japan.* New York; Tokyo: Weatherhill, 1981

Native Industrial Art Laboratory (Hg.): *Shifu Fabric.* Shiroishi: Sendai Print, 1946

Narita, Kiyofusa: *A Life of Ts´ai Lung and Japanese Paper-making.* Tokyo: The Dainihon Press, 1966

Okado, Yuzuru: *Japanese Handicrafts.* Tokyo: Japan Travel Bureau, 1956

Paireau, Francoise: *Papiers Japonais.* Paris: Adam Biro, 1990

Progin, Gisela: *Projekt Shifu. In: Textil Forum Textile.* Basel. 2(1995), S. 12–17

Rathbun, William Jay: *Beyond the Tanabata Bridge. Traditional Japanese Textiles.* Seattle: The Traver Company, 1993

Rein, J.J.: *The Industries of Japan.* London: Hodder & Stoughton, 1889

Rudin, Bo: *Shifu – japanskt papperstyg.* In: Hemslöjden. (1988), S. 26–27

Schulte, Toni: *Kleider aus Papier im Fernen Osten.* In: Papiergeschichte. Darmstadt. 6(1956), S. 43–44

Siegenthaler, Fred: *Paper Webbings. Papier-Gewebe – Papier-Bänder.* Papier-Mitteilung Nr. 41. Basel: Sandoz AG, 1991

Sparke, Penny: *Japanisches Design.* Uda Stätling (Übs.). Braunschweig: Westermann, 1988

Takamatsu, Akemi; Hanson, Jon: *Awagami Factory. Washi: Japanpapier.* Brüssel: C. Masson, 1999

The Shogun Age Exhibiton Executive Committee, Tokyo (Hg.): *Shogun. Kunstschätze und Lebensstil eines japanischen Fürsten der Shogun-Zeit.* Tokyo: Toppan Printing Co., 1984

Watson, William: *The Great Japan Exhibition. Art of the Edo Period 1600–1868.* London: Royal Academy Of Arts, 1981

Weber, Therese: *Kunsthandwerk in Indonesien und Japan.* In: Schweizerische Arbeitslehrerinnen-Zeitung. Zürich: Fretz AG, (1986), S. 1–16

Weber, Therese: *Washi – Vergangenheit und Gegenwart der japanischen Papiermacherkunst.* Basel: Verband der Schweizer Papierhistoriker, 1988

Yamada, Sadami; Ito, Kiyotada: *Handbuch der Papierkunst. Anleitung und Beispiele zum kunsthandwerklichen Arbeiten mit Papier.* Tokyo; San Francisco; New York: Japan Publications Trading Company, 1966

Yoshimoto, Komon (Hg): *Traditional Japanese Small Motif. Textile Design* I. Singapore: Page One Publishing Pte Ltd., 1993

Yoshimoto, Komon (Hg): *Traditional Sarastic. Textil Design IV.* Singapore: Page One Publishing Pte Ltd., 1994

Zyugaku, Bunsyo: *Hand-made paper of Japan.* Tokyo: Board of Tourism industrie, 1942

c) The history of European paper textiles

Ahrens, F. [u. a.] (Hg): *Das Buch der Erfindungen. Gewerbe und Industrien. Verarbeitung der Faserstoffe (Holz-, Papier- und Textilindustrie). (8. Bd.)* Leipzig: Otto Spamer, 1898

Bachinger, Karl; Hemetsberger-Koller, Hildegard; Matis, Herbert: *Grundriss der Österreichischen Sozial- und Wirtschaftsgeschichte von 1848 bis zur Gegenwart.* Wien: ÖBV-Klett-Cotta, 1987

Bauche, Rolf: *Papier macht´s möglich. Geschichten zur Papierverwertung.* (Begleitdokument zur Ausstellung im Rheinischen Industriemuseum Papiermühle Alte Dombach in Bergisch Glattbach) Essen: Klartext-Verlag, 2000

Beha, Willi: *Die Verwendung von Papiergarn in neuerer Zeit.* In: Wochenblatt für Papierfabrikation. Biberach. 68(1937), S. 158–160

Blechschmidt, J.; Englert, L.; Henschel, M. [u. a.]: *Zellstoff Papier.* (5. Aufl.). Leipzig: VEB Fachbuchverlag, 1979

Boelcke, Willi A.: *Deutschland als Welthandelsmacht. 1930–1945.* Stuttgart; Berlin; Köln: Kohlhammer, 1994

Boesch, Hans : *Papierene Kleider im Anfang des 18. Jahrhunderts.* In: Papier-Zeitung. Berlin. 8(1883), S. 540

Deutsches Historisches Museum (Hg.): *Bilder und Zeugnisse der deutschen Geschichte.* Aus den Sammlungen des Deutschen Historischen Museums. Berlin: Deutsches Historisches Museum, 1997

Diels, Ludwig: *Ersatzstoffe aus dem Pflanzenreich. Ein Hilfsbuch zum Erkennen und Verwerten der heimischen Pflanzen für Zwecke der Ernährung und Industrie in Kriegs- und Friedenszeiten.* Stuttgart: Schweizerbart´sche Verlagsbuchhandlung, 1918

Eichholtz, Dietrich: *Geschichte der deutschen Kriegswirtschaft. 1939–1945.* (3. Bd.). Berlin: Akademie-Verlag, 1996

Eigner, Peter; Helige, Andrea (Hg.): *Österreichische Wirtschafts- und Sozialgeschichte im 19. und 20. Jahrhundert.* Wien; München: Christian Brandstätter, 1999

Feldhaus, Franz Maria: *Wer erfand die Papierwäsche?* In: Papier-Welt. Hildburghausen. 2(1925), S. 8–10

Gräber, Ernst: *Die Weberei.* (8. Aufl.). Leipzig: Max Jänecke, 1938

Heinke, Wilhelm: *Handbuch der Papier-Textil-Industrie.* Dresden: O.H. Hörisch, 1919

Hofmann, Carl: diverse Beiträge in: Papier-Zeitung. Berlin. 65(1940a), S. 713ff; 65(1940b), S. 1234; 21(1896a), S. 292; 39(1914a), S. 2080; 21(1896b), S. 1088; 45(1920), S. 2599; 39(1914b), S. 2438

Kuczynski, Jürgen: *Geschichte des Alltags des deutschen Volks. 1918–1945.* (Bd. 5.). (3. Aufl.). Köln: PapyRossa, 1993

Leonhardt, Kurt: *Papierstoffgarne. Zellulongarne nach dem Naß-Spinnverfahren Türk-Issenmann.* Berlin: Akademie-Verlag, 1954

Lion, Leopold: *Die Textilbranchen. Ein Lehr- und Nachschlagebuch für alle Zweige des Textilgewerbes.* Nordhausen: Heinrich Killinger, 1922

Papier auf dem Vormarsch: Vom Brautkleid bis zur Bettwäsche. In: Spandauer Volksblatt. Berlin (1968-05-03)

Papierkleider. In: Wochenblatt für Papierfabrikation. Biberach 6(2875), S. 575–576

Papierkleider: nur eine Frage der Zeit?. In: Der Papiermacher. Heidelberg. 18(1875), S. 86–91

Papierwäsche. In: Wochenblatt für Papierfabrikation. Biberach 46(1915), S. 23

Pennenkamp, Otto: *Über das rationelle Spinnen von Papiergarnen. In: Papier-Zeitung.* Berlin. 69(1944), S. 50f

Petzina, Dieter: *Autokratie im Dritten Reich. Der nationalsozialistische Vierjahresplan.* Stuttgart: Deutsche Verlags-Anstalt, 1968

Pfuhl, E.: *Papierstoffgarne (Zellstoffgarn, Xylolin, Silvalin, Licella). Ihre Herstellung, Eigenschaften und Verwendbarkeit.* Riga: G. Löffler, 1904

Püschel, Erich: *Die Herstellung von starkem Papiergarn. In: Papier-Zeitung.* Berlin. 45(1920), S. 1098

Sultano, Gloria: *Wie geistiges Kokain.... Mode unterm Hakenkreuz.* Wien: Verlag für Gesellschaftskritik, 1995

Sutton, Ann; Sheehan, Diane: *Faszination Weben.* Bern; Stuttgart: Paul Haupt, 1990

Tobler, Friedrich: *Textilersatzstoffe.* Dresden; Leipzig: Globus Verlagsanstalt, 1917

Willinger & Co Europa – Lehrmittel OHG (Hg.): *Fachkunde Textiler Rohstoffe.* Wuppertal: Pittroff, 1996²

d) Current developments

Badisches Landesmuseum (Hg.): *Papier – Art – Fashion. Kunst und Mode. Kleider aus Papier.* Karlsruhe: Badisches Landesmuseum, 1996

Braddock, Sarah E.: *Project Papermoon.* Bech, Claus (Übs.). Kopenhagen: Eks-Skolens Trykkeri ApS., 2000

Callaway, Nicolas (Hg.): *Issey Miyake. Fotografiert von Irving Penn.* Schaffhausen [u. a.]: Edition Stemmle, 1988

Curtis, Lee J.: *Lloyd Loom. Wohnen mit klassischen Korbmöbeln.* München: Mosaik Verlag, 1992

Digel, Marion: *Papermade. Wohnen mit Papier und Karton.* Raumelemente Möbel Leuchten Accessoires. München: Mosaik, 2002

Koumis, Matthew (Hg): *Art textiles of the world.* Japan. Winchester: Telos Art Publishing, 1997

McCarthy, Cara; McQuaid, Matilda: *Structure and Surface. Contemporary Japanese Textiles.* New York: The Museum of Modern Art, 1998

Museum für Kunsthandwerk Frankfurt am Main (Hg.): *Zeitgenössisches deutsches und finnisches Kunsthandwerk.* Schwanheim: Henrich, 1987

Nuno Corporation (Hg): *Boro Boro.* Tokyo: Takeda Printing, 1997a

Nuno Corporation (Hg): *Suké Suké.* Tokyo: Takeda Printing, 1997b

Periäinen, Tapio: *Soul in Design. Finland as an example.* Helsinki: Kirjayhtymä, 1990

Schoeser, Mary: *International Textile Design.* Hong Kong: Laurence King, 1995

Siegenthaler, Fred: *Shifu. Nepalesische Shifu-Herstellung.* Papier-Mitteilung Nr. 40. Basel: Sandoz AG, 1990

Stenros, Anne (Hg.): *Visionen. Das moderne finnische Design.* Jürgen Schielke (Übs.). Helsinki: OTAVA, 1999

Sterk, Beatrijs: *Martin Leuthold und die Schläpfer-Kollektion In: ETN-textilforum.* Hannover: Scherrer. 2(1997), S. 28–29

The Finnish Foreign Trade Association (Hg.): *Design in Finland.* Helsinki: Hämeen Kirjapaino Oy, 1991

Tsuji, Kiyoji: *Fiber Art Japan.* Tokyo: Shinshindo Shuppan, 1994

Wallin, Sirkka (Hg.): *Nirunaru Snoddar & Snören.* Helsinki: Edita Oy, 1997

e) Textile techniques

Barrett, Olivia Elton: *Korbflechten. Vorlagen und Anleitungen.* Augsburg: Augustus Verlag, 1991

Burns, Hilary: *Weiden, Binsen, Peddigrohr.* Bern; Stuttgart; Wien: Verlag Paul Haupt, 2000

Buser, Pia: *Sprang. Eine uralte Technik zum Flechten mit aufgespannten Fäden oder Schnüren.* Arth: Eigenverlag P. Oechslin-Buser, 1980

Collingwood, Peter: *Textile Strukturen.* Bern; Stuttgart; Wien: Verlag Paul Haupt, 1988

Collingwood, Peter: *The Technique of Sprang. Plaiting on Stretched Threads.* London: Faber and Faber, 1974

Crockett, Candace: *Weben mit Brettchen.* Bern; Stuttgart; Wien: Verlag Paul Haupt, 1994

Debétaz-Grünig, Erika: *Web- und Knüpftechniken.* Bonn: Hörnemann, 1978

Deutch, Yvonne: *Weben und Spinnen.* Herrsching: Pawlak, 1985

Dillmont, Thérèse de: *Encyklopädie der weiblichen Handarbeiten.* Mulhouse: Ed. Dillmont, 1908

Fisch, Arline M.: *Textile Techniken in Metall.* Bern; Stuttgart; Wien: Verlag Paul Haupt, 1998

Fuchs, Heidi; Natter, Maria: *Häkeln.* Bassermann, 1997

Georgens, Jan Daniel: *Die Schule der weiblichen Handarbeit.* Berlin: Loewenstein, 1869

Gillow, John; Sentance, Bryan: *Atlas der Textilien.* Bern; Stuttgart; Wien: Verlag Paul Haupt, 1999

Jensen, Elisabeth: *Korbflechten. Das Handbuch.* Bern; Stuttgart; Wien: Verlag Paul Haupt, 1994

Joliet-van den Berg, Magda und Heribert: *Brettchenweben.* Bern; Stuttgart: Verlag Paul Haupt, 1975

Joliet-van den Berg, Magda und Heribert: *Mit Brettchen gewebt.* Brunnen-Reihe, Bd. 116; Freiburg: Christophorus-Verlag, 1976

Klein, Gabriele; Nagler, Maria: *Fadenspiele, Häkeln, Stricken, Nähen, Weben.* Eisenstadt: Rötzer, 1974

Klein, Gabriele; Nagler, Maria: *Häkeln, Stricken, Nähen, Stoffdruck, Mustergestaltung.* Eisenstadt: Rötzer, 1975

Lenz, Charlotte: *Brettchenweben.* Ravensburg: Otto Maier Verlag, 1976

Lundell, Laila: *Das große Webbuch.* Bern; Stuttgart; Wien: Verlag Paul Haupt, 1987

Owen, Rodrick: *Geflochtene Kordeln und Tressen. Ein Anleitungsbuch mit über 250 Mustern.* Bern; Stuttgart; Wien: Verlag Paul Haupt, 1996

Schlabow, Karl: *Die Kunst des Brettchenwebens.* Neumünster: Karl Wachholtz Verlag, 1981

Seiler-Baldinger, Annemarie: *Systematik der Textilen Techniken.* Basler Beiträge zur Ethnologie. Band 32. Basel: Ethnologisches Seminar der Universität und Museum für Völkerkunde. In Kommission bei Wepf & Co. AG Verlag, 1991

Sentance, Bryan: *Atlas der Flechtkunst.* Bern; Stuttgart; Wien: Verlag Paul Haupt, 2001

Sutton, Ann: *Bindungen zum Handweben.* Bern; Stuttgart; Wien: Verlag Paul Haupt, 1987

Sutton, Ann; Sheehan, Diane: *Faszination Weben.* Bern; Stuttgart; Wien: Verlag Paul Haupt, 1990

5.5 Suppliers and Museums

Paper yarn

Fa. Ecofil
Papierausrüstung Alfred Truchseß
(equipment)
Mühlstraße 21a, A-8072 Fernitz
trualf@utanet.at
www.ecofil.at

Papierspinnerei Fa. Julius Glatz
(paper spinning mill - trade)
Talstrasse 268,
D-67434 Neustadt (Weinstraße)
spinning-mil@glatz.de
www.finepaper@glatz.de

Fa. Liebe Dinge
Monika Wagner
Bismarkstraße 10, A-4020 Linz
wagner@liebedinge.at

Fa. Lotteraner KG
Glockenstraße 19, A-1020 Vienna

Fa. Natürlich Pflanzlich
Lars Gather
Anrather Straße 19,
D-47918 Tönisvorst
info@natuerlich-pflanzlich.de
www.natürlich-pflanzlich.de

Fa. Pirkanmaan Kotityö
Pyynikintie 25, FIN-33230 Tampere
anne.lehtinen@ktypirkanmaa.fi
www.ktypirkanmaa.i/varjaamo.html

Fa. Friedrich Traub
Schorndorfer Straße 18,
D-73645 Winterbach
www.traub-wolle.de

Fa. Webkante
Sirpa Lutz-Piiponen
Alte Landstrasse 64,
CH-8708 Männedorf
webkante@bluewin.ch

Fa. Zürcher & Stalder AG
Postfach, CH-3422 Kirchberg

zsag@zsag.ch

Paper for making Shifu

Awagami Factory (The hall of Awa
Japanese Handmade Paper)
141 Kawahigashi, Yamakawa-Cho
Oe-Gun, Tokushima,
779-34 Japan

Fa. Boesner (Japanese and Nepalese
paper)
Brambillagasse 11, A-1110 Vienna
boesner@magnet.at

Fa. Johannes Gerstäcker Verlag GmbH
(Japanese and Nepalese paper)
Vecosstraße 4, D-53783 Eitorf
info@erstaecker.com
www.gerstaecker.de

Fa. Hahnenmühle FineArt GmbH
(Japanese and Nepalese paper)
Hahnenstraße 3, D-37586 Dassel
sven_metze@hahnemuehle.de
www.hahnemuehle.de

Fa. Helvetas (Nepalese paper)
St. Moritzstrasse 15
Postfach 181, CH-8042 Zürich

I. Kessler (Nepalese paper)
Kulturviertel zwischen den Museen
Getreidemarkt 15, A-1060 Vienna
i.kessler@netway.at

Fa. Lempen (Japanese paper)
Mühlentalstrasse 369,
CH-8201 Schaffhausen

Fa. Wolfgang Stadler
(Nepalese paper)
Photografie & Art Paper
Sulzbach 31, A-4820 Bad Ischl
artpaper-photo@utanet.at
www.artpaper-photo.com

Fa. Thalo (Japanese paper)
Soodstrasse 57/59, CH-8134 Adliswil

**Papermaking supplies courses and
museums**

Basler Papiermühle (Basel paper mill)
St. Albantal 37, CH-4052 Basel
www.papiermuseum.ch

Designmuseo- Finnish Museum of Art
and Design
Korkeavuorenkatu 23,
FIN- 00130 Helsinki
www.designmuseum.fi

Deutsches historisches Museum
Zeughaus
Unter den Linden 2, D-10117 Berlin
www.dhm.de

Fa. Eifeltor Mühle
John Gerhard
Auf dem Essig 3, D-53359 Rheinhach
gerhard@eifeltor-muehle.de
www.eifeltor-muehle.de

Österreichisches Papiermacher-
Museum (Austrian Paper Museum)
Museumplatz 1
A-4662 Steyrermühl
www.members.vienna.at/difr/papierm
achermuseum

Papiermuseum Düren
Wallstrasse 2-8, D-52349 Düren
www.papier-museum.de

Papierwespe
Beatrix Mapalagama
Äegidigasse 3/1, A-1060 Vienna
papierwespe@chello.at
www.papierwespe.at

.6 List of artists and companies mentioned in the book

Eja Aro
Nuhmarkoskentie 549,
FIN-62240 Numarkoski
eija.aro@kolumbus.fi

Claudia Bernold-Buol
Alte Landstrasse 253,
CH-8708 Männedorf
claudiabernold@gmx.ch

Ester Chabloz
Rue des deux marchées 14,
CH-1800 Vevey

Anita Dajcar-Florin
Kirchgasse 5, CH-7310 Bad Ragaz
anita.dajcar@bluewin.ch

Renate Egger
Thannrain 25, A-6422 Stams
puntoala@yahoo.com

Mashiko Endo
Takanosu Miyagi, J-Shiroishi-City

Luzia Fleisch
Daney Gasse 1, I-39028 Schlanders
luziafleisch@chrizia.com

Esther Foelmli, Webatelier
Oberdorf 14, CH-4806 Wikon, Affoltern

Katharina Frey
Ritterstrasse 26c, CH-3047 Bremgarten
k.r.frey@bluewinch.ch

Claude & Andrée Frossard
Ronzeru 11, CH-2024 Sauges

Gerlinde Fuchs
Schindlweg 14, A-4161 Ulrichsberg
gerlinde.fuchs@eduhi.at

Marian de Graaff
Karel de Gretelaan 139,
NL-5615 SR Eindhoven
mariandegraaff@hotmail.com

Mona Gustafsson
Kaserngatan 3, S-57535 Eksjö
mona.g-son@telia.com

Asa Haggren
Institute for Konst- och Bindvetenskap
Dicksonsgatan 2, S-40530 Göteborg
asa.haggren@hb.se

Sanja Hanemann
Pfarrplatz 11, A-4020 Linz
sanza@aon.at

Uta Heinemann
Weissenbühlweg 41, CH-3007 Bern
u.heinemann@bluewin.ch

Cordula Hofmann-Molis
Weingartenstrasse 2,
D-83417 Kirchanschöring
adico@t-online.de

Naomi Kobayashi
Kaminaka, Keihoku-cho
Kitakuwada-gun,
Kyoto, 601-0532 Japan
ktsn@fjg.so-net.ne.jp

Hanna Korvela Design Oy
Minna Canthin katu 20–22,
FIN-70100 Kuopio
hanna@hannakorveladesign.fi
www.hannakorveladesign.fi

Sonja Knoll
Lohmatte 3, CH-3184 Wünnewill
s.knoll@freesurf.ch

Korb- und Stuhlflechterei Seestern
Seestrasse 185, CH-8708 Männedorf
flechterei@vspzo.ch

Monika Künti
Werkstatt für Flechtwerk und
Textilkunst
Schifflaube 50, CH-3011 Bern
kuenti@bluewin.ch

Loom GmbH
Justus-von-Liebig-Straße 3,
D-86899 Landsberg am Lech
info@lloyd-loom.de
www.lloyd-loom.de

Sirpa Lutz-Piiponen
Alte Landstrasse 64,
CH-8708 Männedorf
webkante@bluewin.ch
www.webkante.ch

Fa. Ulf Moritz
Studio in Amsterdam
design@ulfmoritz.com
Kontakt in Nürnberg:
Sahco Hesslein
info@sahco-hesslein.com
www.sahco-hesslein.com

Mäti Müller
atelier «papier-t-raummatten»
Ca Menegon,
CH-6545 Landarenca
matimueller@freesurf.ch

Nuno Corporation
Ms. Reiko Sudo
Minato-ku
Axis Bl 5-17-1 Roppongi,
J-Tokyo 106 Japan
nuno@nuno.com,
www.nuno.htm

Fa. Nuno (europäische Zentrale)
Franziska Lüthi
Mittelstrasse 9, Postfach 101,
CH-3000 Bern 26
sain@smile.ch

Projekt Papermoon
Ann Schmidt-Christensen
Projekt Papermoon (Modedesign)
Atelier Plexus
Thorsgade 83. bh. 1,
DK-2200 Kopenhagen

Grethe Wittrock
Projekt Papermoon (Textildesign)
Textil 5C
Norrebogade 5 D 3,
DK-2200 Kopenhagen
grethe.wittrock@post.tele.dk
www.crafts.dk/urban

Veronika Rauchenstein
Aarpark 10, CH-8853 Lachen SZ

Dorothea Rosenstock
Bergstrasse 67, CH-8708 Männedorf

Fa. Ruckstuhl
St. Uranstrasse 21, CH-4901 Langental
ruckstuhl@ruckstuhl.net
www.ruckstuhl.ch

Sadako Sakurai
838-4 Hori-cho,
Mito-shi, Ibaraki-ken, 301 Japan

Fa. Jakob Schlaepfer
Fürstenlandstrasse 99,
CH-9001 St.Gallen
www.jakob-schlaepfer.ch

Marianne Schneider-Lüdi
Mühlerain 25, CH-3210 Kerzers
fschne1953@tiscalinet.ch

Sigrid Schraube
Theodor Heuss-Straße 6,
D-61137 Schöneck
sigridschraube@yahoo.de

Asao Shimura
Kami Philippines Inc.
Tina, Makato
Aklan 5611, Philippinen
asao@my.smart.com.ph

Deepak Raja Shresta
21/400 Putali Sadak,
Katmandu, Nepal
deepak_rs@yahoo.com

Lis Surbeck
Puppikon, CH-9565 Rothenhausen
surbeck-web@bluewin.ch

Teppich-Art-Team
Hugo Zumbühl + Peter Birsfelder
Untere Gasse 1, CH-7012 Felsberg

Textilwerkstatt Bürgerspital
Fr. Monika Cerff and Anita Detwiller
Flughafenstrasse 235,CH-4025 Basel
m.cerff@buespi.ch
a.dettwiler@buespi.ch

Linda Thalmann
Ringstrasse 42, A-4061 Pasching
thalmann@gmx.net
loshase@yahoo.de

Maisa Turunen-Wiklund
Miilutie 21, FIN-65320 Vaasa

Atelier Webzettela Franziska Käser
and Erika Wyss
Belpstrasse 71, CH-3007 Bern
od: Kanonenweg 18, CH-3012 Bern
franziska.kaeser@bluewin.ch

Woodnotes Oy (Ritva Puotila)
Tallberginkatu 1 B,
FIN-00180 Helsinki
woodnotes@woodnotes.fi

Subject index

5.8 Picture index

All the unattributed photos are by Christina Leitner.

The following abbreviations are used: a, above; b, below; m, middle; l, left; r, right
Rolf Bauche, Papier macht's möglich. Geschichten zur Papierverwertung. (Paper makes it possible. Stories about using paper). Document accompanying the exhibition at the Rhein Museum of Industry, Alte Dombach Paper Mill in Bergisch Gladbach,Essen 2000, 8, 30

Anika Galonska, 33, 34, 35
Asa Haagren, 111Ml
Asao Shimura, 141, 143u

Claude and Andrée Frossard, 52
Conrad Schraube, 91r

Deepak Shresta, 145u

Eliane Laubscher, 148, 149a/bl
Erich Leitner, 4, 53 all, 54, 55 all, 58, 78, 83o, 86 all, 95a/m, 97M, 100, 102l / ra, 104b, 113 both, 120 both
Esther Chabloz, 67r, 173

German History Museum Berlin, 32
Gisela Progin, 23 (Dress from Gisela Progin's estate, photographed by Christina Leitner), 47 all, 48 r, 49, 144, 145o, 147, 149br

Hanna Korvela Design, 172

Jens Friis, 164r
Jeppe Gudmundsen-Holmgreen, 165, 166, 167l /r

Katharina Frey, 77ar, 89a
Katharina Frey, 77or, 89o
Keisuke Yoshida, 18
Korb- und Stuhlflechterei Seestern, 107o

Linda Thalmann, 159
Lis Surbeck, 88u, 116
Lloyd Loom, 41

from: Cara McCarty and Matilda McQuaid, Structure and Surface, Contemporary Japanese Textiles, The Museum of Modern Art New York, 1998: Osamu Mita, photographed by Karin Willis, 133

Marian de Graaff, 76 mr, 77al, 162, 163 all
Marianne Grondahl, 164l
Mäti Müller, 150, 151, 153r
Mona Gustafsson,126b, 107a

Naomi Kobayashi, 138, 139, 140,
Nobumitzu Katakura, 22br
Nuno, 175, 176a/b

Renate Egger, 129r

Sahco Hesslein, 174r
Seestern basket and chair weaving factory,
Seki Yoshikuni, 19
Silvia Moos, 135, 136, 165, 166, 167l / r
Sonja Knoll, 68r
    Sukey Hughes, Washi. The world of Japanese Paper, Tokyo, New York, San Francisco 1978, 29

Tapani Pelttari, 168
Teppich-Art-Team, 156, 157, 158 all

Ulla Paakkunainen, 127b

Françoise Paireau, Papier Japonais, Paris 1990, (Japanese paper, Paris 1990). Photographs by Tsunehiro Kobayashi, 18r, 21, 28

Woodnotes, 132, 169, 170 all, 171a/b